CHAPTER ONE

Barrow

THE STORY BEGINS on Tuesday, 6 April 1993. I was a little apprehensive as we walked toward the Town Hall, approaching from the car park. We were two or three minutes early for my two-thirty appointment and as we turned into the entrance a solitary tall man standing to the left caught my eye. He stared quizzically and smiled. Within seconds we were greeting each other. His name was Brian Charles.

My first thought was: What an odd-looking man! It was mainly his head and face, for he wore a good suit, stiff collar and tie. It was windy, so perhaps that was why his top notch of curls looked unkempt. But the colour – red/brown was quite striking. The cut of the rest was "short-back-and-sides" and greying. But it was his teeth, there seemed to be several gaps with solitary teeth here and there. He smiled readily, so the overall first impression was that he *was* an 'oddball.' He carried a brown briefcase, not a particularly good one.

We ushered ourselves into the entrance. Our guest approached the reception desk completely unabashed and asked for the café. Of course, we knew where it was, having checked it out already that morning. Nevertheless, we allowed ourselves to be guided into the area where fortunately there was an empty table. What a relief!

Glancing at the menu, Allan ordered tea and scones all round. Brian Charles said he had just eaten (so had we) but a cup of tea would be welcome. After a time he said to me, "What are you doing here today?" By then Allan had disappeared to pay for the scones and Mark had disappeared shortly afterwards. "Oh, completing a research project with my team. We have been very busy. I'm interested in the Barrow family." He was studying me very intently. I felt uneasy and did not elaborate further.

* * *

The visit had been specially arranged by me to meet this man. I wanted to quiz him about the correspondence he had sent me about my book "The Byrom Collection," published the previous year in 1992. He had heard me talking on the BBC radio programme 'Today' in an interview conducted by John Dunn and it was not long afterward that I received Mr. Charles' first batch of correspondence on 28 November, redirected to me by my publisher Jonathan Cape. Fortunately, my Commissioning Editor, Tom Maschler, had advised me earlier to keep a daily diary and I have followed his wise advice ever since. For that date I recorded:

> I received a strange letter (via Cape) this morning, envelope addressed to 'Joy Hancox (Author)'. The contents cannot be readily understood. And so at this stage I will attempt to record the contents![1]

Little did I know at the time what was in store for me. Turning now to the letter and contents: The envelope, posted from Barrow in Furness, contained a number of pages varying from an introductory letter, an original drawing scripted and coloured with a scenic view, a geometric drawing of the 'Hill of Hoad' in Ulverston and a handwritten piece of philosophical prose. The handwritten letter said the following:

> Dear Lady,
> I have just enjoyed reading your book (The Byrom Collection) and I thought to inform you of the divine tetragram that if it is written vertically it forms

> This could be god himself cosmic man the divine androgyne. Certainly something to superimpose on your Tree of Life etc.
> Best Wishes
> B. Charles
> (Brian Charles)

A strange introductory letter indeed! In addition, the combination of strange pages inside seemed very unusual, even disturbing. The title of the main accompanying feature page 'Song of the Flying Sorcerer,' with the small sketch of a flying saucer at its side, when considered with the letter including the phrase 'cosmic man' seemed to me something of a 'wild card' and beyond anything relevant to my understanding of what I was dealing with at the time. But from time to time my curiosity persuaded me to look at it again. The quality of the presentation suggested that it was meaningful and had an intelligence of its own. I remained curious.

The Messenger

The Messenger

Joy Hancox

© Joy Hancox, 2021

Published by Byrom Projects

A CIP catalogue record for this book is available from the British Library.

ISBN 978-0-9566394-1-7

Book layout and cover design by Clare Brayshaw

Prepared and printed by:

York Publishing Services Ltd
64 Hallfield Road
Layerthorpe
York YO31 7ZQ

Tel: 01904 431213

Website: www.yps-publishing.co.uk

Contents

List of Illustrations

The BYROM COLLECTION

on

Thursday 14 May

at

The Groucho Club

The Gennaro Room

RSVP Nicola Vaughan
20 Vauxhall Bridge Rd
London SW1V 2SA
071 973 9730 x2168

45 Dean Street

London W1

12.30-2.30pm
Drinks &Buffet

Launch Day 14 May 1992

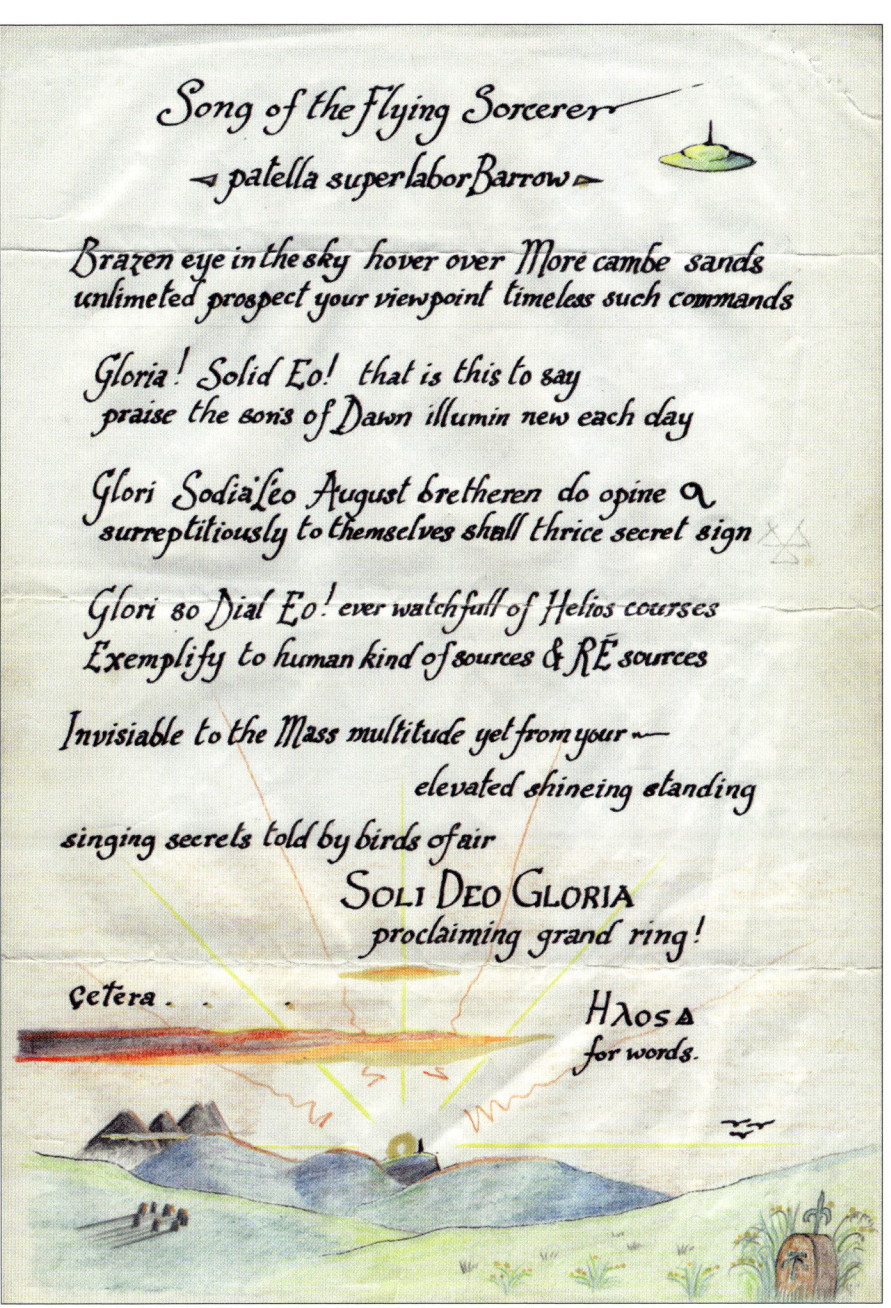

Song of the Flying Sorcerer

"The Byrom Collection" had received some serious attention already over the months, but somehow this correspondence was different. My book was a general introduction to 516 geometric drawings and an interpretation of some – mainly architectural. The architectural drawings I believe are concerned with the iconic design of the Elizabethan playhouses in London, including the famous 'Globe' theatre of Shakespeare fame. For this reason, I was particularly interested in the 'Hill of Hoad' geometric drawing.

Brian Charles had included his address in Barrow, so ten days later a group of us visited Barrow to scout the territory. I wanted to check some of the background to this correspondence and visit the Hill of Hoad and the lighthouse on top. Why had geometric features connected with it been highlighted and sent to me?

We first visited 'Devonshire Buildings,' the address given by Brian Charles. I had no intention of leaving a calling card or meeting him that day. My diary entry for Tuesday, 8th December reads:

We knew exactly where to enter Barrow and as we approached nearby Michaelson Road and the Dock area, it became more and more depressed. I was shocked. We stopped near the 'Devonshire Arms.' Allan went to make enquiries. 'Devonshire Buildings' was immediately behind the pub. There was nowhere to park, so we drove around to find a suitable place. We found a side road near to the Post Office, locked the car door and made our way towards the address of Brian Charles. The road was so shabby and neglected – graffiti on walls, well-worn curtains and washing hanging on lines across the road. I had never been anywhere like it before. '3G' was at the top end, a flat. Allan went inside and up the central stairway. When he returned, he was able to point out the gaunt, curtainless windows of '3G,' the forlorn cactus and wires in and around the windows, together with a bowl leaning up against the sink. We took lots of pictures and might have even captured the dark figure 'thirtyish' at the window.

We took car numbers and more photographs. Next, a visit to the Town Hall and the electoral rolls – no one but Brian Charles at the listed address. We then visited the Civic Centre and purchased paracetamol and a bottle of water, driving away at about 12:45PM towards Ulverston. 'Hoad Hill' is about six miles away. We stopped up a lane at the base of the hill. Only Malcolm, my brother and I walked up it. It was a very hard climb for me. The tower of a hundred feet dedicated to John Barrow (Admiralty) is amazing. Took lots of photographs.[2]

The drive back to Manchester was indeed thought-provoking. Who was Brian Charles, knowledgeable but living in a harsh industrial area and close to the docks? Was he poverty stricken? Yet, he was able to buy my book or did he borrow it? What sort of a mind did he have? Someone able to prepare such papers in such surroundings. I felt a degree of concern and was humbled by his interest. What did he expect of me?

Given the time of year, close to the holidays, I sent a seasonal greeting card as an acknowledgement but did not give my address. My courtesy duty done, I turned my attention to other professional matters and the approaching festive celebrations.

* * *

The completed manuscript of my next book was due with my Editor at Cape at the beginning of the New Year, a biography of John Byrom, an eighteenth-century Manchester man and former owner of the Byrom Collection. The final details were keeping me busy. There was much to-ing and fro-ing of telephone calls and letters. As Christmas rapidly approached my daily routine got even busier and more exacting. Included in my mail on 21 December was another brown envelope, redirected to me by Cape from this correspondent in Barrow.

I really had not had time to study his first correspondence. I found the contents of this second envelope paradoxical. There were two separate sheets; the first was an arithmetical computer diagram issued by Elstree Aerodrome in connection with Airtour products. On the page Mr. Charles thanked me for the 'card.' He was sending me "herewith my Yuletide greetings."

He seemed to be associating himself with the airport at Elstree. Was he a pilot or technical boffin? I recalled taking photographs of an aeroplane circling quite low overhead round the Hill of Hoad as Malcolm and I were scrambling up. That was curious since we were the only people about at the time. My only other thought in connection with Elstree was of the famous film studios nearby. Fanciful thoughts. I moved on to the 'greetings.' They were handwritten on a second sheet. Eighteen lines of nine couplets. They begin:

Twentieth of December the old year breaths (sic) his final icy
blast he gave us good and bad times now his power is failing fast.
Twenty-second of December a Mighty gulp of breath a whole
brand new year in front of us, no time to think of death.

Reading the lines through, I felt rather confused. Any connection between the pages escaped me, although he had suggested that the computer diagram might be associated with a piece of equipment I had referred to in my earlier book. Hardly so. I felt that this was an excuse for including other information on the sheet which *he* wanted me to know. The motif at the top of the greeting intrigued me. Clear, bold and hand-drawn, the Knights Hospitaller of St. John's cross! So meaningful to me. The correspondence would have to be put to one side. What was the motive of the sender? I had visited the area three weeks earlier out of curiosity. I telephoned Barrow library to see whether they had a copy of my book. They did not.

The manuscript of my new book "The Queen's Chameleon" was collected by courier on 15 January and delivered to Jonathan Cape the next day. But the day before, I received yet another letter from Brian Charles, redirected from Cape. Who was this man?

In an effort to deal with this pre-Christmas correspondence from him, I sent him a suggested reading list of 'Hermetic' books, including a study of Robert Fludd[3], the seventeenth century astrologer and mathematician. I still had not been able to decide whether there was a proper theme to the letters. The topics seemed miscellaneous and not straightforward. In fact, I detected that the subject under discussion might well be codified. I had no wish for there to be a continuous flow of letters, which Cape dutifully and very efficiently was forwarding to me. At this point I had not met Mr. Charles.

In any case, the next few weeks were to be fully occupied with finalizing the draft of my new book with my editors. I informed Mr. Charles of this and suggested that when I was next in the Barrow area for research purposes, we could perhaps meet to discuss the correspondence and I could sign his copy of my book.

So once the pressure on me had become less, I arranged with my research team to visit Barrow on 6 April. A meeting with Mr. Charles was arranged for 2:30PM. It was against the backdrop of my earlier visit to the area and the impact of the bleak surroundings of Devonshire Buildings and the memory of my arduous climb to the top of the Hill of Hoad still fresh, that I was anxious to obtain some answers to what I had provisionally concluded was a set of well thought-out, industrious, but practically incomprehensible set of correspondence.

We were a team of four. Each member had been well prepared beforehand. So there were five of us at the round table with our light refreshments of tea and scones. I sat next to Mr. Charles who I gauged was in his mid-fifties. I wanted to hear what he had to say about the 'Hill of Hoad' geometric drawing. I referred to what I believed was Masonic symbolism in the drawing and asked whether he was a Freemason. He said that he was not, but that he had read a lot of 'their' books. He seemed more interested in telling me about his own visit and experience inside the tower at the top of the hill. I asked whether he had used some sort of code when constructing the pages. He said simply "No" but that they were "very deep." He talked of the history of Tarot Cards and Elaine talked with him about Tarot Readings, with which he seemed familiar. I asked whether he had any family. He said that he had four children, but now he was very much "a loner." He seemed reluctant to talk in any sort of detail about the rest of his correspondence and I found his steady gaze in my direction most disconcerting. Allan came to my rescue at one point, engaging in topics connected with everyday matters. I then asked whether he worked locally. He answered vaguely, saying he had worked on the motorways as an engineer but was now retired. I signed his copy of my book.

An hour had passed and Mark reminded me of my next appointment. (A tactful arrangement made earlier.) Immediately, Mr. Charles opened his brief case and took out an envelope and handed it to me. A cursory glance showed more pages of the same kind, which I was not going to read in front of him. He then gave me a leather bookmark stamped with gold lettering 'Furness Abbey.' We shook hands and I thanked him. Allan escorted him to the door of the café. Then he was gone. We sat for a few minutes, silently reflecting on what had been a truly strange encounter. Soon we were driving out of Barrow just before 4:00PM, each of us mulling over the questions I had posed at the meeting. Once back in Manchester, we continued our discussion over a Chinese supper. Without much hesitation, Allan and Mark gave their views: Mr. Charles had had a nervous breakdown, since generally his responses were limited and unsatisfactory. Elaine was remarkably hesitant and non-committal. As for myself, I was rather disappointed in one way, but most of all so preoccupied with his strange appearance that I decided to draw a sketch of him before the end of the day, dwelling on his odd features as I remembered them. I wanted to be able to recall the details, as we had taken no photographs at the meeting.

The day had been long and it was not until the next morning, Wednesday, that I read the pages from the envelope given to me by Brian Charles. I had glanced at the mail waiting for me on my return, but my curiosity persuaded me to examine the contents of the envelope first. There were four pages of what seemed to be verses, entitled:

Hymn of Furness Abbey

I must admit that at the end of my first reading I was quietly impressed. The piece consisted of one hundred and two lines and signed:

Fra Furness Abbe

But my next reaction was accompanied by my recollection of Brian Charles. It was almost immediate. *He* was not responsible for creating the piece I had just read. Before looking at it again I placed the pages back in the envelope and decided to check the unopened mail from the previous day. To my surprise there was a letter from Sam Wanamaker, the American actor and impresario in charge of building the new 'Globe' theatre project taking place at the time on Bankside, overlooking the Thames in Southwark.

The letter was an invitation to the Preview of 'The Merry Wives of Windsor' on 22 April. It was to be performed by a German company at the now partly completed 'Globe.' Only two of the twenty bays were standing at the time. The rest of the structure was simply the general frame. I felt that my invitation was probably due to correspondence I had been having with Theo Crosby, the architect in charge of the project. We had been having a dialogue about the design since 12 June 1989, three and half years previously, and although the academic forum of advisers was less than enthusiastic about my interpretation of the original Elizabethan design, Theo Crosby had written to me several times about it. In one note, on 8 January, he wrote about his latest plans for the Globe in which he said:

> You will see I have borrowed your Byrom Collection drawing
> and fused it somewhat with what we have been given from
> other sources. I hope you will approve.[4]

With all this in mind I accepted the two tickets for the German company's performance on the 22 April. A step forward I thought.

Having made the necessary arrangements, I had a little time to spare that morning and being curious, I returned to the enigmatic Barrow envelope. Neatly scripted, with additional small drawings and phrases of an explanatory nature added strategically, the verses I realised, *were* a coded message of an allegorical nature. It was smart and structured and I sensed there was a real purpose behind the 'giving.' There were 102 lines and it was when I came to line 91 that my awareness became even more focused. From here it reads:

> Could we some clement time parlay
> On pointers from the olden day Then
> from the viewpoint of days of yore
> See the promised land by golden shore
> Two leagues from where I sit today
> I'll show to you this grand display
> Conjure fools and sages, many maids and men.

Study what they gained and lost, make them live
again. Ears to hear have you, and eyes that seek to
see. Then let us spin a yarn or two –
If not let me be.
Fra. Furness Abbe.

At this point I could not have understood the earlier part of the piece, but from line 91 onward the intention was fairly obvious. Remembering the man who gave me the envelope and Barrow itself, I became even more confused. I decided that the suggested meeting proposed in the verse was not for me. I would not proceed further and did not. Someone else was behind Mr. Charles and his communications to me. It seemed as though he was a messenger of sorts, acting in the role of 'postman.' His whole demeanour suggested that he was not capable of such a presentation.

So, my Barrow episode was over. That is what I believed at the time.

The Globe. Two Bays

CHAPTER TWO
'The Merry Wives of Windsor'
Preview 22 April 1993

ON 8 JANUARY 1993 I received from Theo Crosby, the architect with Pentagram Architects in charge of the 'new' Globe theatre design, his latest drawing. I had sent him my observations regarding it. By then I had become a 'player' involved, independently on the side, but recognised as having at least some expertise. On receipt of the invitation from Sam Wanamaker to attend a play there, I would now be visiting the actual site in two weeks' time. The play was 'The Merry Wives of Windsor,' to be performed in German by The Bremen Shakespeare Company.

It was ironic that in my manuscript of the John Byrom biography, which I had recently submitted to 'Cape,' I had written of Byrom himself going to see the same play on 2 March, 1724, some two hundred and sixty-nine years before. Where would it have been performed? Byrom does not say. His diary suggests it was near Leicester Fields in London. There was no Globe theatre standing in the city at the time.

I was looking forward to the occasion and although attendance was by invitation only, I managed to get an extra ticket for my London agent, Anne Dewe. We travelled by taxi from Anne's office in Soho to Bankside and made our way to the Museum. My diary entry for the day reads as follows:

Had any instructions been left for us? No! So we made our way to the Globe site. A corrugated fence was still intact. A few people were ambling this way and that. We walked towards a gap in the fence and quickly on to a makeshift Bridge. This is what it seemed we had to do, but nobody told us. There was no one there to greet us or to check our names. We could have been anyone. Once inside the arena I quickly looked around this bleak spectacle. There were interesting company members purposely involved in attempting to reach some sort of deadline. A few 'hangers on' were huddled under the small covered area. The rest of the auditorium was open to the skies and simply-tiered, structured. An odd body here and there occupied a space, but mostly it was empty. We bantered and jostled verbally as to where we should sit (there was ample choice). Cameras seemed to be part of the uniform of most of the visitors, for they were in the main media men and women and journalists. Too soon I was looking towards the stage area. A toy-like curtain was stretched from one side to the other – 'tawdry' as a description is rather charitable I think. It was easy to see it was a collection of stained and different coloured curtains sewn together. Against the ill-ordered setting and surroundings, it looked far too amateur. Painted men in silly, shabby costumes (1930s) pranced around the 'groundlings' area, soon distributing German programmes. A few extra people arrived. Allan counted 45. It was all very odd and very cold. A bit of a 'warm-up' next – audience participation – a few laughs. Sam Wanamaker ambled into the opposite side. Sat on his own. Perhaps it was a journalist who joined him? He looked bent and frail and with a 'bloom' of impending old age. Sad, solitary, with a camera poised every so often. The play started. I could not understand any of it. That is all I can say. I am sure the Germans would have understood. But visually it was so crude I did not particularly want to understand. The actors were audible but there was no atmosphere. The tawdriness was broken only by a noisy helicopter and the occasional engines of the cruisers on the Thames and the aircraft carrying traffic to hopefully warmer climes. We shivered and shivered up to half time (end of Act Two). Our English programmes arrived just before the interval. We decided to leave. A German TV Journalist stopped us dead in our tracks. "Are you leaving? "Yes", we replied. "Why are you leaving?" "It is very cold. Laurence Olivier himself couldn't prevent me from leaving. It is so cold on the 'open' benches. I wonder how Shakespeare is being received in German by an English audience?" "But you, I think, are about the only English people here."

I told her that my particular interest was in the design of the theatre and, since it was in the state it was, I had seen enough for my purposes. It was no reflection on the German company. It was very cold. I did mention the curtains – suggested the Laura Ashley panel should be in a different position and that a five minutes break could have been helpful to an English audience. 75 minutes was a very long time to sit to attention listening to it all in German. I wanted to break away from her but felt extremely guilty for feeling as I did. I really ought to stay, I thought, but there was Anne who was anxious to go and Allan who was turning purple. I too was cold but tried to appease my own conscience by moving as if to speak to Sam Wanamaker. He was engaged in a less than enthusiastic dialogue with 'Falstaff.' I decided not to speak to S.W. at the point where I heard "It is all very, very difficult" coming from 'Falstaff' to S.W. His facial expression said everything. And so that was the Preview![1]

The event was surely supposed to be a celebratory occasion of William Shakespeare's birthday on 23 April. The performance was to be the next day. It took place on the site, then in development, of what was to be a replica of Shakespeare's Globe theatre. Barry Day, a member of the Globe theatre's board of directors, recalls in his book "This Wooden O" the reception the event received in the press:

> The Times with its headline *Wilkommen and Welcome All* caught the moment best. The groundlings stood or squatted; those seated on bare planks huddled under umbrellas, the multi-coloured traverse drapes whisked back and forth, and the actors successfully competed with planes, helicopters and the chat of the Globe attendants. Despite weather and VIPs it was an oddly homespun, intimate occasion, funny and rather moving. That, too, perhaps, was rather English.

Sadly, eight months after this event Sam Wanamaker, the American inspiration for the project died, his passion and dream yet unfulfilled. The Preview occasion was itself a reflection of the jaundiced view of Culture and the Performing Arts in Britain fashionable at the time in officialdom. I could not understand why I had been invited, me being just a bit player. But then I reminded myself that the German TV journalist had said that we were "about the only English people" present. It was then that I understood and felt isolated. Was this why I had been invited by Wanamaker himself? Although I had noticed his frailty at the preview, I could not have known how ill he was.

My dialogue with Theo Crosby about the Globe design continued and on 11 June I received a major package from Pentagram – 14 pages of architectural drawings and matters related, which had to be addressed in my thinking. These included features connected with audience 'sight lines.'[2] I gave my undivided attention to the matter. Theo and I carried on our exchanges over what I believe was a serious design problem in the proposals envisioned by those responsible for the decision-making at the time.

By 23 August Mrs. Marina Blodget, Sam's personal assistant, was writing to my agent to say that he was expecting to be in his office on 13 September and "we will contact you after his return to schedule a meeting." He was in California at the time. We never had that meeting. The new 'Globe' had by now become something of a beleaguered endeavour with disparate views and insufficient funding. The death of such a charismatic figure as Sam Wanamaker was a great loss. How would the phoenix rise from these ashes? The project simply had to continue.

The first months of 1994 were taken up with the final preparations for my book "The Queen's Chameleon." The launch of the book took place on 21 April at John Byrom's birthplace in Manchester, now 'The Wellington Inn,' a popular hostelry favoured by tourists. Once the preliminaries of the launch were over, I found myself back in contact with Theo Crosby, concentrating on Globe business. Theo himself had been taken ill with an aneurism in February, but nevertheless resumed work on design features from home. On 9 May 1994 I wrote to Theo as follows:

> Dear Theo, I was so sorry to learn of your recent illness and am delighted to hear through your secretary that you are well on the road to recovery. As you will understand, I have been reluctant to intrude and write to your office while you were not available, but I have been following all Globe developments with keen interest. Now that your secretary assures me that you are receiving correspondence, I am writing to you today. I am wondering how far you have been able to follow your intentions – as in the architectural plan section FFGG JJ KK drawn by K.B. referred to as "J (Jan '93 Globe revised)" and incorporate/fuse the Byrom Collection drawing into the latest proposed design of the Stage House. If you recall, we had some later discussion by letter on drawings concerning sight lines etc. copies of which you sent to me. Perhaps, when you feel able, you will let me know the latest position.
> I continue to offer any help that seems appropriate.
> With all good wishes,
> Joy Hancox

Theo replied the next day, 10 May. I include his response (written in his own hand), out of respect for the professional camaraderie we shared at this time:

> Dear Joy,
> Thank you for your letter. After 3 months I am recovering satisfactorily and am much involved with the Globe stage house at the moment. We have the auditorium structure now completed and it will be thatched, plastered and fitted out over the next year. The designs for the stage house structure are now finished after a long tussle between academics, archaeologists and actors/directors. The final scheme represents the best compromise I have been able to get accepted and I am now finalising the details of the frons scenae. It's a delicate business as the period is just on the cusp of a stylistic change to high classic. To do a Jacobean detail now is a great mental challenge. We asked Jon Greenfield to send you the latest drawings. I've not checked them against your book except in the most casual way. Our concerns have been so immediate and often based on personal confrontations. Let me have your comments.
> With good wishes from Theo.[3]

On 16 June Theo telephoned to say that a set of the latest architectural proposals for the theatre design would be sent to me over the next few days and again he would be grateful for my comments. As always, I replied promptly with my considered opinion. On 8 September I sent a request of my own to Theo. There had been so much discussion between us over the last five years concerning the Byrom theatre drawings that I felt a facsimile of the whole collection of 516 drawings in book form was desirable for collective scrutiny. I felt he might be able to suggest a synthesis. I did not receive a reply. Theo died four days later. The shock for me was enormous. As my diary records:

> I did regard him as one of my real supporters and had high hopes with my latest initiative. I really must find a way through despite this sadness. At least now I will feel free to quote from his letters to me – something of the essence of his frustrations over the Globe. The days activities were clouded by the vacuum his passing had caused. I will have to re-adjust again.[4]

Despite the fact that it was thought that he was on the way to making a full recovery, by some quirk of Fate he was knocked down by a cyclist, had a relapse and died on 12 September. So, what was to happen next? With the two main-springs gone from the engine, how could the project move on? Move on it did. Theo had predicted in his letter of 10 May that by the summer of 1995 the auditorium would be "thatched, plastered and fitted out." And so it was! Jon Greenfield, Theo's assistant and successor as architect to the Globe project played a major role in achieving that aim.

Having out of courtesy allowed some time to elapse, I wrote to Jon Greenfield early in December to ask whether my last letter to Theo had surfaced. He replied on 8 December, suggesting that I might write to Polly Hope, Theo's widow. "She won't mind at all." If she could assist she would. On 8 January 1995 she replied, having just returned from New York, "to face a mountain of mail." Despite her willingness to help I decided the matter would be best left alone as far as she was concerned.

Nevertheless, it was time for me to reflect on whether Theo considered the theatre drawings which I had recognised as part of the Byrom collection were indeed worthy of his attention. I remembered his enthusiastic response at our first meeting in the London offices of Pentagram on 12 June 1989, when he saw the original drawings I had taken to show him. I can still recall the initial reaction of this experienced and much respected architect. On that day he summoned his fellow architects from their desks to circle round the table to share his excitement.

Surely, he wouldn't have done so if he had not recognised certain 'truths' within the Collection. He, above all understood my own carefully considered interpretations and intentions at the time. But he had difficulties generated by his responsibilities to Sam Wanamaker. The policy was that he could not move on with any part of the design without approval from the Shakespeare scholars on the Globe's Academic Board. The Board was headed by Andrew Gurr of Reading University and John Orrell, a

Canadian from Alberta University. This had been the case for several years before I had appeared in Theo's office.

The pressure on those left to deal with the unfinished Globe project after the deaths of Wanamaker and Crosby can only be imagined. I was not a member of any committee and was not privy to the decision-making process already in place or those steps that now needed to be taken. Any positive thoughts about my dialogue with Theo could have been known by his successor, Jon Greenfield, but he had now to be the architectural "fixer" for the Academic Board and was charged with bringing the project to its completion. No easy task.

I had long recognised that Theo understood a certain 'restraint' was evident among the esteemed university scholars to anything I had to say. They had been immersed for many years in searching out 'facts' for this design of a replica of the Globe. After all, despite archaeology and ground penetrating radar on the site of the original Globe and a few random historic papers and the panoramic views of London done by Elizabethan and seventeenth century artists, their collective initiatives over the years had not produced the evidence they had hoped for.

The Byrom theatre drawings presented in my book "The Byrom Collection" were actual and real architectural plans. I had been able to commission the model-maker of Manchester University's School of Architecture, Kenneth Peacock, to make models of six of the Elizabethan theatres according to the designs I had interpreted from the drawings. I had not given Mr. Peacock the names of the theatres in my 'Brief.' Each was given a number to make it a 'blind' test and the designs were confirmed by the experiment.

The uniqueness of these designs was not mine. I was merely an interpreter. Furthermore, the models were completed before the publication of the book. It must have been difficult for the Academic Board to accept that such evidence existed at all. When "The Byrom Collection" was launched in London on 14 May 1992 at Groucho's, I decided to invite Theo. From my theatre archive I reproduce the following correspondence:

28th April, 1992
Dear Theo,
"The Byrom Collection." I have asked my agent to send you a copy of my book by special courier today. I hope it has arrived. I am sure you will find it easier to read than the Galley Proof. The illustrations are certainly much better. I have tried to include a fair representation of the drawings and still hope to include them all in a facsimile edition one day.
I would like to invite you to a lunch-time launch at Groucho's on 14 May. Cape will be sending out invitations in a day or so and I have asked them to include you. If you are unable to come I will understand, but I would very

much like you to be there so that I can thank you personally for the interest you showed in the early days. [1989] The Times Saturday Review will be publishing an article on 9 May.
Best wishes, as always,
Joy

Theo's response was:

Dear Joy,
Thank you very much for your book which came today. I shall go over it very carefully and be truly grateful. I don't think I can come to the Groucho Club, but I will be with you in spirit.
Best wishes,
Yours,
Theo[5]

The launch at the Groucho Club was an occasion to remember, with representatives from the literary world and the media. There were also representatives from the London English Heritage offices who had shown interest in the Collection for some time. The event culminated in a book signing at Hatchard's. It was a day of celebration – but no Theo Crosby. He clearly thought his attendance would send out signals not appropriate for the 'Globe personnel' of the moment. Even so we continued to correspond, exchanging ideas and data. Considering Theo's legacy afresh, his letter of 14 January, 1993 (which was sent a few weeks before 'The Merry Wives of Windsor' performance) is the one which encourages me to recognise the stance he took with regard to the Collection of theatre drawings and my work. He saw the situation clearly and understood, despite the difficulties. The letter makes this plain:

Dear Joy,
I was delighted, not to say overwhelmed, by your letter and enclosures. I've gone over them carefully and I think you have made many useful points. I'll have to sweat it out as usual. May I say how grateful I am for the trouble you have taken.
With best wishes,
Yours,
Theo[6]

But others seemed to take a different view. Despite the very rich collection of skilled professionals drawn from business, politics and academia, no one at this juncture chose to follow up Theo's interest. The uncertainties all round were masked by anxieties. The need to consider the possibilities within the geometry of the Collection and its historic background seemed inappropriate and unnecessary to them. The cursory interest expressed by a few in the Globe organisation had become latent.

For me at the time the void was bridged by matters arising from my latest book, "The Queen's Chameleon," the Byrom biography, and problems engulfing the Globe project were of necessity pushed into the distance. My own beliefs remained secure, enveloped in the unique numerology (mathematics) involved and the exquisite precision of the 'setting out plans' of the geometry in the drawings and my licence to deal with them.

CHAPTER THREE

Anomalies

A S AN INDEPENDENT researcher working on material that had not been available to mainstream scholars, I recognised that the Byrom drawings created understandable difficulties for those scholars. In truth it seemed as though I was not welcome in "their" workplace, so I continued to look for more evidence and into their 'provenance' (origens) for years to come. What lay behind the elegance and beauty of the Collection?

Not one of the 516 drawings could be later than 1732. Again, I used forgotten and unpublished material belonging to John Byrom to gain a better understanding, which I then included in my biography of him, published the same year Theo Crosby died, 1994. I pointed the reader in a direction which shows how Byrom's philosophy had led him to acquire such a collection. His visits to Bartholomew Close in London to purchase the papers of Mr Faulkner, alias Mr. Rose, on hermetic interests in 1735 provided a clue. As already mentioned in that publication, Mr. Faulkner was a member of the De Bry dynasty of printers and publishers responsible for some of the Byrom collection of drawings.

In 1605 the Rector of the nearby St. Bartholomew's Church was David Dee. That year he was forced to flee from his 'living' because of religious differences with his 'superiors.' He was, I believe, a kinsman of John Dee, the famous mathematician, astrologer and occultist, who was also forced to abandon the wardenship of the Collegiate Church (now Cathedral) in Manchester the same year, perhaps for similar reasons.

John Dee is the creator of the Elizabethan theatre designs and must have known members of the De Bry family. While the Globe was being built Dee was out of Manchester. He was Warden of the church there from 1595 to 1605. Records show that one Rowland Dee was a 'butcher' in St. Bartholomew's Close in the seventeenth century. I believe that this Rowland Dee was John Dee's son. I also believe that John Dee stayed in the area just across the river from the construction site during the building of the original Globe in 1599, when Dee was recorded to be out of Manchester.

John Byrom was born in Manchester and died in Manchester. He was very familiar with John Dee's legacy. Byrom's library of books shows that he was an admirer of Dee's remarkable collection of writings. My sporadic accumulation of this data helped me to remain steadfast to my beliefs regarding the matter. Unfortunately, the Globe academic scholars chose not to discuss such possibilities as planning for the new 'Globe' proceeded.

However, as my diary entry for 23 January 1995 shows, each day brings with it a part of what becomes life's rich pattern of events:

> There was a very interesting letter in the post this morning from Mark Rylance the young Shakespearian actor. I had heard of him some months ago from Leon C. He has a 'Tree of Life' in his dressing room. He was full of praise of 'The Byrom Collection' and would like me to participate in the Shakespeare Birthday Celebrations – 23 April – as his guest. I was very curious since he says that he is a member of the Artistic Committee, Bankside. He has written to Andrew Gurr, requesting a reason for not taking the drawings into account in the design of the 'Globe'. He has received no reply.[1]

After various considerations and phone calls, it was arranged that a meeting should be held here in Manchester on 14 February. Mark Rylance would bring his wife Claire van Campen. He also requested that the Francis Bacon biographer, Peter Dawkins, and Mrs. Dawkins should be present. Mr. Dawkins was also an architect. That is how he was introduced to me. In arranging the meeting, I was anxious to give them the opportunity to read Theo Crosby's correspondence and study the original drawings, of which I was the Licensee.

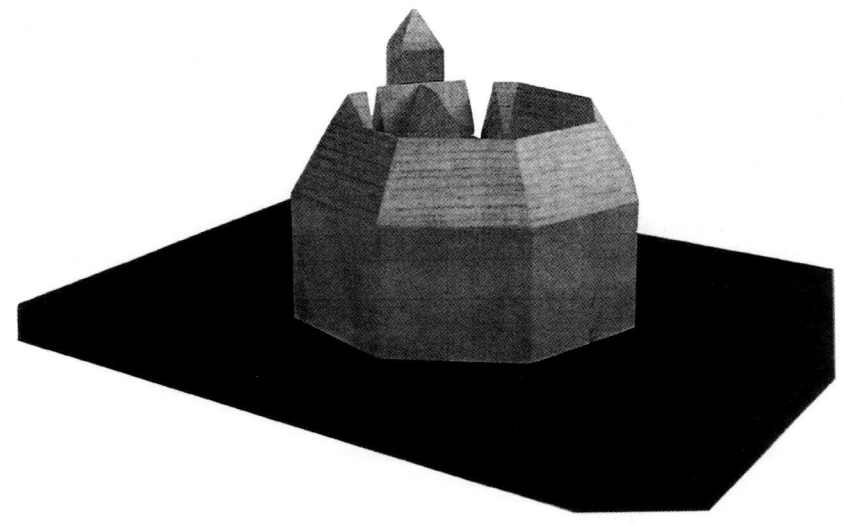

The Globe. Geometry Wooden Model.

THE GLOBE

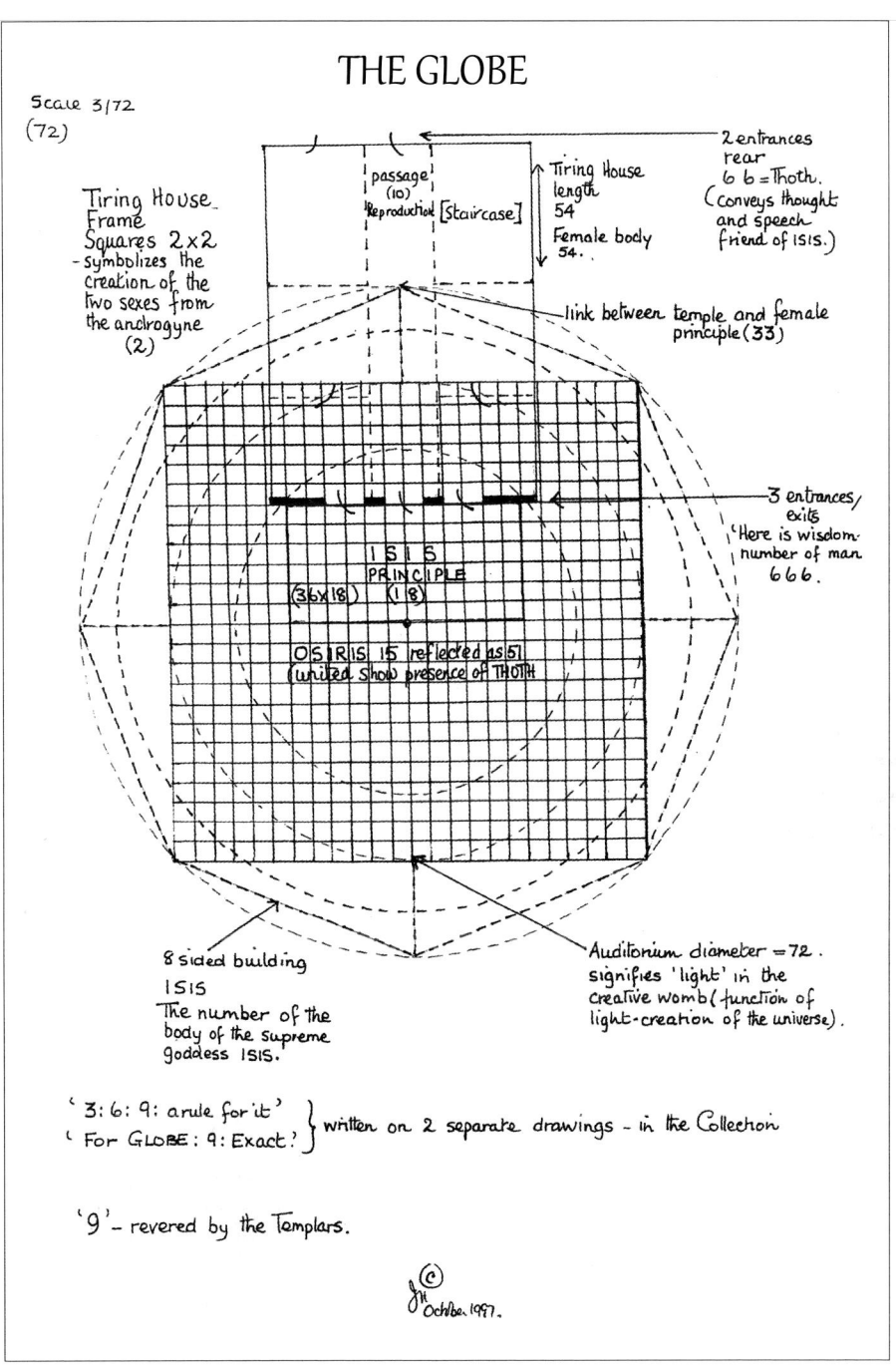

Scale 3/72
(72)

Tiring House
Frame
Squares 2×2
- symbolizes the
creation of the
two sexes from
the androgyne
(2)

passage
(10)
Reproduction [Staircase]

Tiring House
length
54
Female body
54.

2 entrances
rear
6 6 = Thoth.
(conveys thought
and speech
friend of ISIS.)

link between temple and female
principle (33)

ISIS
PRINCIPLE
(36×18)   (18)

OSIRIS 15 reflected as 51
(united show presence of THOTH

3 entrances/
exits
'Here is wisdom.
number of man
6 6 6.

8 sided building
ISIS
The number of the
body of the supreme
goddess ISIS.

Auditorium diameter = 72.
signifies 'light' in the
creative womb (function of
light-creation of the universe).

'3: 6: 9: a rule for it'
'For GLOBE: 9: Exact?' } written on 2 separate drawings - in the Collection

'9' - revered by the Templars.

©
Jtt
October 1997.

The Globe layout 3:6:9 – A rule for itt

21

The day was certainly one that will remain in my memory. As soon as they arrived at about 12:00 midday, the excitement of my intellectual journey about to be witnessed and shared was tangible. There was no disguising the delight they showed in studying the original drawings rather than the published copies. We spent some time examining together the '72 ruler' I had devised to help me understand the numbering system applied by the creator(s) of the collection. In addition, they had the opportunity to study the model of the Globe I had participated in making, according to my interpretation of the drawings. Later they read the collection of letters sent to me by Theo Crosby together with my responses. It was a very emotional time for me. I will never forget their reactions to the few hours we were together. For the occasion, I had also invited those who had been part of my journey of research thus far. Our guests left an hour before midnight.

As a result of the visit, Mark Rylance decided to recount his views of the day to the Artistic Board of the Globe enterprise under the chairmanship of 'Lord Birkett.' In a spirit of good will I decided to accept the offer Mark had made sometime earlier to be a guest-presenter for the first 'Spearshaker Lecture' at the Globe Education Centre on 22 April.

The session needed to be memorable and 'caring.' Preparing for such a venture was indeed a challenge for me. It was an opportunity to demonstrate how I had come to recognise what I believed to be true about the drawings – as architectural plans for the Elizabethan theatre scene of the sixteenth century. After all, I had been working with the drawings for eleven years by then, since 1984. During the next few weeks I prepared booklets for each member of the prospective audience, selecting the appropriate drawings depicting the basic measurements and construction of the Globe. I also prepared individual '72 measure' rulers to go with each booklet. I wanted it to be a 'hands on' session, albeit for one seemingly short event.

It had to be effective. I had no idea beforehand who would make up the audience. With two members of my research team, I travelled down from Manchester a day early and we were met at Euston Station by Mark Rylance. He drove us to the Education Centre where we were introduced to Michael Holden, the Chief Executive of the institution. He was with one or two colleagues and apologised for not being able to be present the next day, as he was due to be at another commemorative Shakespeare event in Stratford upon Avon. We spent a short time assessing the facilities available for the next day.

Saturday was a wet day, so Mark's celebratory commemorative 'walk' in the morning was hampered by the weather. I arrived at the Education Centre, otherwise known as The Bear Gardens, with my team in the early

Globe. A viewing: Work in progress

First Spearshaker lecture. 22 April 1995

afternoon only to find the doors locked. We parked the boxes containing our precious artefacts, including my models of the theatres, on the wet pavement. Very soon conscientious volunteers arrived to open up and help prepare for the expected audience. Among these helpers were Lucy Beevor and Patrick Spottiswoode. Having completed his walk, Mark Rylance appeared promptly at three o'clock and with his customary aplomb, introduced me to the audience.

The small theatre was full and the members of the audience were keen to inspect the explanatory notes in the booklet I had prepared. Showing how the ruler is the key to the measurements of the original Globe's dimensions was the highlight of the afternoon for me, especially since the models were there to be used as visual aids. It was very exciting to be given the opportunity to display the results of my work and research. I had hopes that my efforts were not to be wasted.

The time passed all too quickly. Suddenly my allotted time was over. I waived any fee and expenses. It was all for a good cause and I was very pleased to have had the opportunity to share my findings with a live audience, courtesy of Mark Rylance's invitation. His appreciation was demonstrated by giving my team a dinner that evening with his family at Luigi's restaurant, Covent Garden.

Perhaps the occasion can be best summarised by the letter sent to me by Lucy Beevor, the Public Relations Manager, three days later, on 25 April:

> Dear Joy,
> Spearshaker Lecture. A thousand thanks for all that you did for Saturday's event. I heard nothing but appreciation from the audience. Please pass on our thanks, too, to Allan, Elaine and Mark. They were stalwarts. I do hope the rest of your weekend in London went well. Do come down again next year and join in Mark's walk from Westminster Abbey. Once again, many thanks for a wonderful afternoon. It was a pleasure to meet you.
> Kind regards,
> Lucy Beevor
> Public Relations Manager. International Shakespeare Globe Centre Ltd.[2]

In participating with Mark Rylance in the first Spearshaker Lecture I felt that I had gone to the very heart of where it was all happening. At last I had been involved in a live interaction with an audience. Perhaps now, six years after my first meeting with Theo, there would be a channel available through which the knowledge within the Collection could be shared.

Unfortunately, for the rest of the year, apart from a few congratulatory cards received immediately after the lecture from a few private members of the audience, there was an eerie silence. Behind the scenes there must have been great activity. For me the silence was broken by one important fact. After an interview on Monday, 19 June 1995, Mark Rylance was appointed the first Artistic Director of the Globe, a position he held for the next ten years. The following year, 1996, was a very different year. My diary entry for 5 January begins:

> Post. A card from Mark Rylance. It reads "Well. Well. Well. Here we are in 1996. I start on Tuesday. We have to decide by February. A real Heart and Header. – Mark Rylance." I am glad that he has written. What a job he has! And what is he expecting me to do?[3]

The decision he was referring to was the final design of the Globe in all its aspects. The partially completed new 'Globe' was to be a 20-sided building. My interpretation was an 8-sided building, as is clearly documented in J.C. Visscher's engraving of London dated 1616. Visscher, an eminent printer of his day, had worked for Michel le Blon, a family member of the De Bry dynasty and his initials were on the back of some of the original drawings in the Collection.

In my view, therefore, the building already in progress was inaccurate. It did not fit the geometric pattern dictated by the Pythagorean number symbolism which I had recognised in Byrom's drawings. My interpretation demonstrated that the theatre was more intimate, based on a circular frame of 72 feet in diameter with an outer 'walkway' passage set in an outer wall (8-sided) of 102 feet in diameter (avg.). The stage performing area was based on proportional measurements to be set against the 'tiring house' (backstage). It was an impressive building based on the proportions of the Biblical 'Ark of the Covenant.'

Theo Crosby had told me in his letter of 8 June 1993 that he had somehow incorporated the 'tiring house' from my drawing into his own plans, but I had not seen the evidence of that so far. It was therefore with a continued sense of isolation that I contemplated Mark Rylance's note at the beginning of 1996.

Still, my isolation was not total. We need to return to 1995, for I had received a letter dated 9 May from America. It was from Professor Emeritus John Gleason of San Francisco University. I quote from his letter:

> I am able to demonstrate that contrary to what is regularly asserted Claus Janszoon Visscher, engraver as you remember of the 1616 Londinium, did indeed visit London in 1606. This means of course that Visscher is an eyewitness to what the Globe looked like and what he says he saw, he saw.[4]

I received this letter four weeks after I had given the Spearshaker Lecture with Mark Rylance. For some time, prominent Shakespeare scholars had disputed Visscher's accuracy in his depiction of the Globe in his engraving. That had been a consideration in the then-current academic discussion regarding the design of the new 'Globe.' (I place the name in quotes because it is not the original Globe.) I realised that my eight-sided Globe (again, as depicted by Visscher) could be regarded as an accurate representation of what he had seen and recorded. However, enthusiasm for my views were minimal. The academics no doubt had continued with their own assumptions as they were being expressed in a structure, soon to be seen emerging on London's Bankside. I began a serious correspondence with Professor Gleason, which resulted in a visit by him (together with an architect-colleague of his) to my home on 29 December 1995 to study the original drawings.

Yet even before that, on 8 November, I had a short private visit from Jon Greenfield, Theo Crosby's successor architect associated with Pentagram, and Peter McCurdy, the Master Carpenter in charge of building the new Globe. Also present was my model maker, Kenneth Peacock, who knew Greenfield from his student days in Manchester. It was not a very agreeable session. Copies of Professor Gleason's thesis were shared in confidence, but it could be seen that the die had already been cast for the new design. The practicalities for any change would be enormous. Time and money were a premium. They returned to London late afternoon.

Before leaving for his flight back to San Francisco on 15 February, Professor Gleason had a meeting with Jon Greenfield on 1 February. The meeting lasted for one and a half hours and during a phone call to me later Gleason said it was not without 'tension.' He had nothing to lose and so was frank about Visscher. He said Greenfield 'was all but dismissive' about the Visscher issue.

From his base in San Francisco Professor Gleason continued with his research programme in preparation for his own book. He wrote to me on relevant issues and phoned regularly on matters that needed some discussion. It became clear that he agreed with my views of the original Globe design. Given his thesis on Visscher, this added weight to my seemingly unorthodox quest.

Despite a very busy schedule following his New Year card, Mark Rylance made a courtesy telephone call on Sunday, 31 March. I recorded the main strands of the conversation in my diary. It reads as follows:

> Mark's call was long. It's clear my work on the Collection was considered to be too late, so they say. And if they are 50-100 years late, the drawings are still relevant. (What an admission!) You will have an opportunity in 5 years' time, when the design will be reviewed and [the theatre] *'possibly rebuilt.'* Other Globes will be built and perhaps 'yours'! I will come and work with you, he said. (Dear! Dear!). He is sending some material re: 23 April. I really should try to go. I told him of Gleason's contact and correspondence. [Emphasis added][5]

Mark's light-hearted banter belied the seriousness of what lay behind the call. I wondered at the time who was behind such a declaration! It told me so much. His reference to the 23 April (1996) was in connection with the second Spearshaker Lecture. Claire van Kampen, an accomplished musician and composer, had been engaged to lecture on Michael Maier (1568-1622), sharing the platform with Noel Cobb, the philosopher and author. I thought it prudent that I go and did. It was held on Saturday, 20 April and it could only be a day's commitment. My husband, Allan, had just come out of hospital after undergoing major heart surgery. I was accompanied by one of my research assistants on the trip down to London. On reflection, I think my diary entry best records the day:

> The Blueline taxi service was here to collect Elaine and me at 8:40AM. We boarded the 9:30AM (airline seats) at Manchester Piccadilly for Euston. The train was full of would-be marathon walkers. It was a good-humoured carriage. The train was late into Euston about one hour. Faults of various kinds along the route caused delays. I rang Allan (fine) and caught a taxi straight to Bankside. The roads were not at all busy and our driver enjoyed the run as much as we did. He had not visited the new Globe before. It was a lovely day. Because we were short of time we went straight into the 'Globe.' I explained that I wanted to visit the bookshop and see the stage. There was no charge. How delighted I was to see 'The Byrom Collection' in such a prominent place in the bookshop, next in fact to 'Theatre of the World' by Frances Yates. I had no idea. We took photographs of course for posterity. The stage is no better. The pillars are almost grotesque. My views have not and will not change. I left £5 in the 'Globe Collection' area. From there we walked to the Thames and joined the visitors to watch the Masques. Interesting and certainly a buzz about the place. However, the actors/actresses, though beautifully and authentically dressed with wonderful masques, were over indulgent in their interpretations (I thought). We moved on, had a cup of tea, sitting at a pavement table and then to the Education Centre. The doors were not open. Claire and Natasha (carrying sandwiches) walked down the pavement. We exchanged greetings. It was nice. "Oh! If I'd known you were coming, I would have been better prepared," Claire in a slight panic. We sat at the front, at the side. There was the same frenzy as last year, minutes before. Claire and 4 singers selected 8 fugues of Michael Maier. He wrote 50 when he was 50. Claire gave a lucid and easy commentary and the singers were competent. She welcomed me and mentioned last year's lecture. Coffee break. Noel Cobb's second half was again easy, conversational and informative (salt, mercury and sulphur). I spoke but briefly to Mark and acknowledged Noel Cobb. Mark's mother (Anne Waters) came to talk and spoke warmly about last year. Lunch sometime? Interesting woman. All very pleasant. We walked to the Underground, passing Southwark Cathedral. Visitors were being given a single red rose. We would have loved to stay. At Euston I rang Allan again

[fine]. Julian, (our son) had been ringing through the day. The train was full, including some foul-mouthed football supporters and a school party. The former really let themselves down. The journey seemed very long because of them. Elaine and I ate our prepared sandwiches on the train, going and coming back. Our plans went smoothly. A taxi from Piccadilly took us back to 'No. 11.' Over a cup of tea we gave Allan a résumé of Claire's lecture and the day.[6]

Michael Maier was for some years an important member of the court of King James 1st and his daughter, Elizabeth of the Palatinate. Born in Germany, he was a physician as well as a composer, also a Rosicrucian. The ideals of Rosicrucianism include the reconciliation of the various world religions. In the Byrom collection there are geometric drawings with Maier's name written on them. One of the most significant drawings has commemorative dates placed on one side of a triangle – the date of his death being one. I have identified the geometry in this drawing with important locations and features in Westminster Abbey. They show that certain locations in the layout of the building (internally) are geometrically significant and meaningful, but only to those who have an understanding of the geometry.

In Claire's exposition on one selection, one of Maier's 50 fugues in 'Atalanta Fugiens,' she identified important musical features in its structure with a unique interpretation and understanding. It seems that hidden formulas in musical harmonic structures and the numeracy within the structure of buildings can convey historical and philosophical 'truths,' Michael Maier being an exponent of such skills.

During the next weeks and months, Claire and I continued to share our acquired knowledge on this enigmatic figure, Micheal Maier. It seemed to me at the time that Mark (together with Claire) was anxious to include the awareness of 'hermetic' studies of the Elizabethan period and beyond in the educational studies regarding the Globe theatre. The recent appearance of "The Byrom Collection" in the bookshop was evidence of that. Mark's newly appointed role as Artistic Director encouraged me to think so.

To become familiar with the principles of Hermeticism, one must become aware of the origins and conventions of many diverse cultures and their historical traditions. They are based on the laws of nature and the Universe, incorporated into their religious beliefs. The 'NewAge' practices developed in the 1970s may be considered to be echoing some of the same principles. However, the more orthodox individuals of whatever creed sometimes regard 'New Age' devotees as crankish and gullible, if not eccentric.

Every generation has its own fashionable thoughts. In our own time, with the more recent developments in technology, and considering all the sciences and media, the spread of change is sometimes overwhelming. I have found that in moments of silence and stillness past memories have a chance to enrich the human psyche. It was at just such a moment that I recognised the '72' measure as units of '8' in the Globe design pattern. It was as simple as that. I think my early training as a musician helped me in that.

My journey thereafter took me into the world of Hermetic studies. I became very protective of the fact that the knowledge I gained was part of the construction of the geometry involved. There is an integrity supporting hidden truths. Lines, dots and circles have been used in numbering systems throughout the ages. The Elizabethan, John Dee, used the principles of Pythagorean geometry in designing the sequence of playhouses in the capital. They were in effect replicas of temples and were to be seen in London over a period of approximately fifty years. And then they were gone, yet another casualty of the Civil War and religious conflict.

The theatres were originally built thanks to the patronage of wealthy landowners, some of whom were patrons of the acting companies designated to perform Shakespeare's legendary plays and the works of other writers, such as Ben Jonson and Christopher Marlowe. All acted under licence from Elizabeth and later James 1st, through the office of the Master of the Revels. The Globe theatre was only one playhouse, although probably the most famous at the time. The American Sam Wanamaker's aspiration to house another Globe exactly as the original was highly significant.

According to Mark Rylance's telephone call earlier on 31 March, the drawings had come too late for them to be considered. The design was settled in principle. Yet the Globe bookshop had "The Byrom Collection" available for sale. A paradox indeed. Since my essential interest and concern was with the conceptual understanding behind the design of the Elizabethan theatres, I felt that it was inappropriate for me to intrude any further into the progress otherwise taking place during the next months. After all, knowledge of my research had been available since 1989. We were now enjoying the summer of 1996 and the main structure of the new theatre was well on the way to being completed. My interpretation played no part of that.

The whole project seemed to me to have become more of an 'Enterprise,' with a developing protocol. This dictated that there would be limitations for 'outsiders.' Certainly, my own research programme was gathering its own momentum. It therefore became inevitable that I had begun to distance myself from the 'Bankside' development.

It was with some surprise that I read the contents of a letter which arrived on Saturday, 20 July from a Globe representative. My diary entry for that day records the following:

The post brought me a very interesting letter this morning – an invitation to the 'Stage and Staging' Conference. (The Prologue Season Review on September 8th, a Sunday.) The invitation is from Mark Rylance and Rosalind King. I will give some thought about it anon.[7]

After serious consideration I made my decision that to attend the Conference was a 'must,' but I would need to have been present at some warmup performances beforehand and witness the new theatre in use. I would attend two performances of 'Two Gentlemen of Verona' on consecutive evenings, sitting in two different positions.

So, on Thursday, 5 September, I made what was for me my first historic visit to the now all but completed Globe. My diary entry for the occasion records my honest impression.

The journey across London was difficult. Armed with a packet of sandwiches each and a drink for the interval, we arrived to join a long queue. We sat in the first gallery (stage right) at the side. It was a wonderful experience, despite the design errors and bad acoustics. (Everyone sounded flat.) Mark, though, managed to overcome it – a very moving performance. Dennis the dog stole the show. Our hired cushions helped our comfort. 'The Two Gentlemen of Verona' in modern dress had worked, audience participation – immediate. The theatre could be a success, but whether the novelty will wear off is the question. The human voice is not reproduced at its best. There is a strain which must be tiring for the actors in the long run. These are real issues behind the Shakespearian 'mask.' We will sit somewhere different tomorrow. We went back to the hotel on the Underground – uneventful but quite easy. A memorable day.[8]

The next day, Friday 6 September, was another visit to the Globe – to the same production but arranged differently. My records reflect:

Our journey over to the Globe was easier. We decided to get over there before the rush hour got under way. We had tea and banana cake from the little café opposite the Globe and stood 'Number 1' in the queue. A bearded busker entertained us pathetically. Claire, Mark, Natasha, elder sister and friend went to the café, saw us. We had a lovely pre-play chat (10 mins). We sat ground level, back row (not so good a position) sound often muffled and incoherent. The lights were intrusive, blinding in places. My immediate neighbours put on sunglasses. High spot after the show – we walked with Dennis the Dog and owner. We have enjoyed the day so much again.[9]

Sunday, 8 September, was Conference Day, best described as I recorded it myself at the time. It was a full and busy day:

After breakfast I decided I had better get a taxi and not risk being late. We were there very quickly and we sat for a while on the Embankment. Soon I was joining the delegates in the foyer of the Globe. Allan left. Tea and coffee were waiting. I introduced myself to Rosalind King. It was all a bit 'shilly-shally' for a while, each of us making what you will of the surroundings and stage. Gradually the numbers grew to about 25. Huddles merged in the 'Stars Mall.' I remained on my own. I didn't know anyone sufficiently well. As the deadline for the hard stuff agenda loomed closer, Mark and Rosalind became locked in a personal debate close to the stage. This encounter went on for about 15 minutes at least. We were invited to sit in a side bay – conference table and microphones erect. Preliminaries over, Jon Greenfield talked of the development of the stage area. Questions were invited. It was all to be recorded. I asked a question about the tip of the stage canopy cavities within, trapping the sound – acoustics need to be looked at by an expert. This problem (apart from the overall design) is for me the main stumbling block. It will have to be faced. During the trial scene runs directors intervened, gave their ideas, argued, bickered and bantered. It was not an easy day. I had a word with Jon, said 'Hello' to Hildy. Later Gurr was favoured with a 'Good afternoon.' 'Oh! I didn't know you were going to be here!' (Am I some insignificant nothing in Gurr's eyes? Yes! I think so!) Costume design/colours/textures. More tea! Finish and then the play again, this time centre front, acoustics flat, worse than ever. But it was pleasurable watching the audience. Dinner afterwards. Thank goodness Allan was there. Michael Holden sat opposite me – he didn't have to – we were stuck with each other the whole evening. We both coped quite well I think! Gradually the verbal agenda got round to the drawings. Mark R. joined us. We left at 11:20PM. Dear! Dear! Almost locked in the Underground. Dinner very high powered indeed – all on the 'Globe.' I wouldn't have missed it, but then I had to be there. I would never have another opportunity. What next?"[10]

It was a time for reflection. I had made but a small contribution at the conference itself, but it was a valid one and was recorded. I felt that the dinner was of more significance, in that sitting opposite to Michael Holden, the Chief Executive of the Project, would make our interchanges memorable for both of us. Or so I thought. I did not envy him or those close to him the task ahead. The next months would be critical.

Completely unexpectedly, exactly two weeks later I had a phone call mid-afternoon on 22 September from Mark Rylance. My diary entry for that date reads:

He (Mark) said the Artistic Directorate were unhappy about the 'Globe' stage and Frons/Tiring House. He wants to put the idea of a temporary stage forward (and even try the Visscher, now that there is fresh evidence etc). Would I be prepared to go to the meeting on Tuesday next – 10:30AM?

Mark said he hadn't been able to reach John Gleason. I rang John Gleason (USA) straight away (8:30AM their time) 4:30PM. He is not amused, not to have heard from Mark, and isn't happy for his findings to be presented. I phoned Mark and advised him to phone John G. which he did straight away. They had a long talk. After three quarters of an hour I phoned J.G. back. He said he had offered to fly over in a week or so to present his paper (at 'Globe's' expense). I thought that I would have a little breathing space myself. The shock of this group of calls took my breath away. How could this possibly have happened? What is going on? Surely they would not take the new 'Globe' down at this late stage – it seemed almost unthinkable. I was really very anxious and restless for the rest of the day.[11]

The next day Claire, Mark's wife, rang me up. He was in a meeting. She reported that he would really like me to attend the London meeting the next day. He would phone me a little later. I already had a busy schedule planned for Tuesday in Manchester. George Murcell, the actor and director, was to have a meeting with me to discuss his plans for his own theatre. I cancelled it and went to the Globe meeting in London instead. Mark Rylance was then the newly appointed Artistic Director and his urgent request for me to be present persuaded me that I might make a useful contribution after all. I was to have my meeting with George Murcell later. On Tuesday we caught the 6:45AM Pullman train to London. My diary captures the day as it happened:

If there was a prize for difficult days, I think that this day would get on the short list quite easily. The train journey was uneventful. The queue (Euston) was the longest I've ever seen and a drunk jumping the queue at 9:15AM because his son was dying in Lancashire didn't help. He was disconcerting and potentially violent. We arrived at Bankside just before 10:00AM (Allan left, as usual). It didn't seem that I was expected and I was kept waiting some few minutes before Mark appeared. We walked across to the Globe. He was button-holed by some female. I didn't speak to him again before the meeting or during it for that matter. Rosalind King showed me the agenda. I was not amused. There were about 15 at the meeting including late comers. Andrew Gurr chaired the meeting, but there was little actual formality and no one introduced me. Clearly, I was the only unfamiliar member. I had no idea of the 'content' of this meeting or who in fact the group were. It was the Academic Board! Mark put forward the 'Artistic Board's' dissatisfaction with the performing area and the suggestion that the stage should remain temporary. Jon Greenfield, architect, and Peter McCurdy, builder, were not amused. The suggestion was quite impractical and there would be no opening next June. The permanent fixture/plan should go ahead and *be reviewed in 1, 2, 5 years*. Gurr had prefixed the meeting by saying that all major decisions had been made. There really was no place for me here, but I did say my bit about fresh evidence having been available (for change) since

1992. John Gleason's forthcoming book. There was no response except from Jon Greenfield acknowledging that this was the sort of evidence we/ they were looking for. The rest of the meeting was devoted to decoration. There was little or no regard for performers. I spoke on their behalf – an appeal for consideration – I was but brief. At the end of the meeting after a short chat with the costume manager I told Mark that Andy Gurr and 'their' programme was secure. I left. I sat on the pier after, eating a cheese sandwich and made notes. Allan arrived at 3:00PM. We had to get a taxi to the station – £12 (Euston). There had been a bomb scare. We had a fish supper at Euston, the train was late. It was a long ride home. [Emphasis added][12]

It seemed clear to me that the hastily arranged meeting was in vain. I am not saying that it was a complete waste of time because the practicalities of a completion date were made clear by Peter McCurdy. Any interruption in work on the building programme would be a severe complication. Shortly afterwards it was announced that H.M. the Queen would celebrate the official opening of the Globe, with H.R.H. Prince Philip, on Thursday, 12 June 1997. So the sense of direction was influenced by finishing the job in hand. Anything I said might be seen as an irritation to those under pressure and that was understandable. As I saw it, Professor Gleason's work on Claus Visscher would fall into the same category. My delayed meeting with director George Murcell took place on 8 and 9 October and progress was made in the theatre design proposals for his project, as interpreted from some of the Byrom drawings. An application for Lottery funding was in hand for that theatre as well.

CHAPTER FOUR

Cultural Confusions 1996

B Y SEPTEMBER 1996 it was indeed settled. The Wanamaker London 'Globe' was to be a twenty-sided black and white thatched building. But that had really been decided in 1992, before the publication of "The Byrom Collection." I had made the information contained in the drawings available to the architects in 1989 and by 1992 two of the twenty bays had been erected. I did not know what had happened 'behind the scene' at the time that had been responsible for such an accelerated building programme. Still, I had no entitlement. My book was launched in May of 1992.

After many years of concentrated study, I was certain that the 1599 construction of the Globe was eight-sided as noted, and all else followed from that. I did not understand how the decision for something else had been made so quickly and at the time was not aware of the financial dilemma surrounding the project. I was an independent researcher. Later, I was to learn that a business transaction had occurred between two parties: the Globe administration and 'another,' to the mutual benefit of both. An exchange, all above-board and proper, meant that timing was compressed. Jon Greenfield and Peter McCurdy were to play a major role in the arrangement. Theo Crosby and Sam Wanamaker must have been subscribers as well, with the academics Andrew Gurr and John Orrell giving the nod. Doubtless, each of these had his own team of supporters. The transaction meant that two bays went up almost immediately and once they could be seen on the London skyline the design was a given. Historical and cultural niceties became a detail at that point.

The years between 1992 and 1996 had witnessed the deaths of two distinguished pioneers of the venture, Sam Wanamaker and Theo Crosby. They had both been sympathetic to my project interests. To others, it became patently clear that I and my work on the 'Collection' and what it meant for the Globe project was an embarrassment. I did not fit in and could not be carried along. My presence at the hastily arranged meeting on 22 September 1996 made it clear that I and my input were superfluous and I reconciled myself to the present set of circumstances. The new 'Globe' would not be a true replica of the 1599 building. Did it really matter, some would say? To me it did. Everything to do with the Globe was centred round William Shakespeare and his legacy of plays and sonnets. To quote Francis Bacon:

> Sometimes Truth is so painful
> it must be buried and forgotten.
> Twisted, misused. Specially by
> those in Possession of Power.[1]

Historical uncertainties have provided intrigues for centuries.

* * *

I was becoming increasingly busy with my two recent books in the public domain. Contacts and correspondence were a constant. Regardless of the 'closed shop' atmosphere now engulfing the Globe programme, the topic did not seem to want to go away for me. In my mail for the day, after my meeting with George Murcell on 10 October, there was yet another letter from Mark Rylance dated 7 October. It was accompanied by a graphic showing the results of a recent radar scan of the original Globe site at Anchor Terrace carried out by the Museum of London. The survey was expected to provide further information to the GPR investigation which had been carried out earlier, particularly in connection with the stage area.

Despite understanding that the new Globe was to be different from the interpretation I had established, and if therefore spending precious time on this latest scan would be a waste, my curiosity was alerted. I found myself engrossed with the drawings, my special ruler along-side them. Courtesy persuaded me to reply to Mark on the 19 October. I quote from my response:[2]

19th October, 1996
Dear Mark,
Thank you for sending me a copy of the Globe ground plan with the latest archaeological radar scan superimposed on it. Naturally, I found it of great interest and have spent some time studying it alongside my own interpretation of the Byrom drawings.
It seems to me that if you re-position the stage etc. in the direction which I have always maintained is the correct one, i.e. N. /N.E. in keeping with the Visscher engraving, then the numerous dots which indicate 'soil disturbance' appear to lie naturally along a series of concentric circles. In other words, although at this stage we cannot be sure what these dots represent, a pattern can be seen in them which agrees with the circular shape of the theatre and is more meaningful than in the arrangement of the Globe with its present orientation. In diagram 'C' I have similarly moved the theatre slightly to

the right to show that the Hancox/Byrom plans fit into the configuration (but facing the opposite direction, as with the Visscher).

Would you please show the diagrams to Michael Holden? Perhaps it would be useful (and logical) if I could have the opportunity to share my findings with those responsible for the present radar scan. Could that be arranged?

I did not receive a reply from Mark until Christmas Eve, 24 December. His response contained a bombshell of a shock which became a preoccupation over the season's festivities. I include a complete copy of the letter which was typed on Globe headed note paper. The consequences became all consuming:

20 December 1996

Dear Joy,

Thanks for your letter in October. I am sorry about the delay and about not being able to see you the other day. Just too busy.

I am in bed with the flu tonight and hence undisturbed and able to read and study your diagrams. I just do not feel that these 'dots' give us enough to go on. I see what you are trying to do but until we have something more concrete or less concrete…by a couple of feet….isn't it incredible to think the answers are so near! I will pass on your drawings to Michael Holden but I do not think there is enough evidence to move the debate forward. **I was amazed to hear that an individual has bought all copies of The Byrom Collection and it is now out of print.** Do you know who it is? A Merry Christmas to you and Alan. Did you know that mount Parnassus had two peaks, is this the reason for the double gable on the second Globe!?

Love from a many-sided round place. Mark.

P.S. We have set up dates for a Globe Review of Evidence from within and without the Globe. I hope you will be able to attend. I will get you an agenda! Preliminary dates are Thurs 4/9/97 – Sun 7/9/97. Might try to get Prof. Gleason to come.[3]

It was the first I had heard about all copies of "The Byrom Collection" being bought up. The letter arrived on Christmas Eve so I could not get in touch with the publisher or my agent to find out whether it was true. There were other aspects of the letter which caused confusion. Using my diary entries to guide me, I have noted here my cause for concern.

Some weeks earlier, before receiving Mark's letter, on Saturday, 26 October, I had received an invitation from Jon Greenfield, the Globe architect, inviting me to a follow-up meeting to the previous September 'conference' on the following Wednesday, 30 October. It was to discuss the stage area again. By that time, I had written my letter to Mark Rylance and had received no reply. I decided that there would be no positive or realistic contribution I could make regarding the stage at this juncture, so

I wrote to tell Greenfield that I would not be attending the meeting. Two days after that conference, on Friday 1 November, still having received no reply to my request for a possible meeting with those responsible for the latest radar scan, I decided to take matters into my own hands. I spoke to a representative of the Museum of London on the phone who gave me the name of the contact I needed, Dr. Bill McCann. I spoke to Dr. McCann that afternoon and after some discussion it was agreed that we should meet in his office on Tuesday, 12 November at 2 o'clock.

We travelled by train to London Euston and then by Underground to Queen Victoria Street. My diary for the day reads as follows:

> Walker House was just a short way. My hat blew off, the umbrella was turned inside out, all in a space of seconds. What a start! Dr. McCann met us on Floor 3 – pleasant, unassuming, short, Irish. His assistant, Paul, made us a cup of tea. I was soon recounting the Mark Rylance episode (radar drawing). I showed them Mark's copy of the drawing as an introductory note and my reply to Mark's letter requesting the name of the 'body' in charge of the 'exploration' and the opportunity to meet. No reply etc. When Dr. McCann and Assistant saw the drawing(s) they were absolutely shocked.[4]

Dr. McCann said he didn't recognise the findings upon which I had superimposed my own drawing as his own up-to-date research. It had been altered. I was taken aback. Michael Holden had commissioned the work in November 1995 and it had been completed by March 1996. There had been some 'in-house' discussion about the results for the Report of the findings which had delayed its printing, and when it was printed this is what I had been provided with. It was with regret that I found myself privy to the disquiet of these two professionals sitting across the table.

There was obvious confusion that needed to be resolved. I promised to consult Mark Rylance about these 'findings,' which he had personally sent to me, as soon as possible. My husband telephoned Mark's office shortly after the meeting ended. He was told that Mark only had a half hour slot available the following morning at 9:30AM. We booked it. But shortly afterwards on the same day, reflecting on what I thought might turn out to be a serious issue which needed more time for consideration, I rang up and left a message cancelling the interview. In any case, I wanted to reflect on what I had heard that afternoon.

Despite further efforts to contact Mark Rylance during the next days and weeks, I heard nothing until his letter dated 20 December, received on that Christmas Eve. In the meantime, however, I had been shown a tentative interest on the part of Michael Holden in the possibility of the publication of a Facsimile Edition of the complete collection of the

Byrom drawings. It quickly came to nothing – too costly but desirable, it was said. There had also been further contact from Dr. McCann on 15 November. A colleague of his from Sidney, Australia wanted to get in touch with me as a result of our meeting on 12 November. I felt that was a little premature. There was much to consider. Something was definitely amiss with respect to all these developments. I told Dr. McCann that I would wait to hear from Mark Rylance. His 20 December letter was the response I was waiting for.

The Christmas and New Year holiday period had meant it had not been possible to obtain facts regarding Mark Rylance's comment on the purchase of the entire available Byrom Collection books by 'someone,' until 9 January 1997. Having received a memo on that day from my agent, I telephoned the newly appointed Managing Director of Cape, Dan Franklin. He was able to confirm that the sale had indeed taken place 'last spring' (1996). The book was now out of print, as was my later book "The Queen's Chameleon."

No one from the firm had chosen to inform my agent or me. My Cape editor, Tony Colwell, was completely unaware of the situation as well. I was unable to discover who the individual purchaser was and have never been able to, but it was apparent that for some reason the books should not continue to be 'in the public domain.' This became more clear a short time later when it was announced that the actual negative of the book had been lost. Butler and Tanner had been the designated printer for Cape and they could not locate it. This was now 1997, during the months leading up to the grand opening of the new 'Globe.'

It became something of an embarrassment for a while because a 'misunderstanding' between publisher and author had occurred. Jonathan Cape is a prestigious publisher. Fortunately, a gracious understanding was achieved by Dan Franklin. On 19 March it was agreed that a paper-back version of "The Byrom Collection" should be published. The hard-back copy of the book would need to be scanned because of the missing negative.

Publishers reserve the right to terminate contract agreements, making sure that conditions on the author's behalf have been met. The decision to publish a paper-back edition of the book within the next few months speaks for itself. Clearly, there were two issues I had to consider: the embarrassing issue in connection with my earlier meeting with Dr. McCann in his office at Walker House at the headquarters of The Museum of London and now the supposed disappearance of the stock of "The Byrom Collection" books and the missing negative of the book by the printers, Butler and Tanner. Dan Franklin's decision regarding the introduction of a paper-back version went some way to dull my disappointment and curiosity at the time. In any case the 'paper-back project' kept me occupied for the following weeks, which went on to become months.

And yet, there was that 'stinging' reminder from Dan Franklin's initial phone call on the 9 January that the decision regarding the sale of the books had been made "in the spring" of 1996. He would not have recognised the significance of the spring reference to me. As it so happened it was on the 31 March 1996 that Mark Rylance had phoned to say that the drawings had come too late to be considered in the design phase of planning the Globe. Then, much later, during our meeting at Walker House, Dr. McCann explained that his commission regarding the interpretation of the Radar Scan of the Globe's original site had been completed by March 1996. He told me that on 12 November. So 'spring' 1996 had been significant on both counts. I felt that my efforts were being intentionally supressed and that it was 'spring' of that year when certain decisions had been made by certain individuals in an effort to do so.

Coincidentally, it was Mark himself, whose letter I received that Christmas Eve of 1996, who told me that the books were bought up by an individual and he did not seem to know when or by whom. I reminded myself that detailed measurements of the Globe drawings and the overall interpretation of the design were published in my book, which differed from the new Globe design as set out by Jon Greenfield and the Academic Board.

It may well have been a straightforward policy so far as Jonathan Cape was concerned, but I do not consider it too fanciful to think that there was a connection between the two events somewhere along the way. However, I got on with the job in hand, which was supplying new illustrations for the paper-back edition because the scanned versions were not good enough. The facts in connection with what was for me the timing of this disturbing episode cannot be denied. But any correlation linking any of it together must be speculation because I was not privy to the hard facts of any meetings that took place. It is so easy to make guesses and then speculate. But the reality of actual events and the timing can also present core truths. I continued in blind faith at the time, noting each day's events in my diary.

The circumstances surrounding my biography of John Byrom, entitled "The Queen's Chameleon," were a different matter. The subject of the book had nothing to do with the Globe project as such. Although the supply was bought out at the same time as "The Byrom Collection," Dan Franklin was able to arrange 200 courtesy copies to be sent to me and the accumulation of all these decisions became part of a developing mystery. I was beginning to suspect that not only the 'hidden' understandings behind the design of the Globe, but all my work was being suppressed.

While I have always known that there are those who feel that 'hermetic' knowledge should be kept secret because 'the world isn't ready for it,' I find that hard to believe 'in this day and age.' Is it all a coincidence? Providence? Human folly? "Perhapes Shakespeare Knoweth the Truth," someone I know would say, maybe Francis Bacon.

CHAPTER FIVE

'1997'

THE FIRST MONTHS of 1997 remain some of the most mystifying of my life. In the powerful world of publishing my two Byrom books had been consigned to silence – without my knowledge! Apart from 'tit-bits' now and then, nobody in or around that world wanted to talk to me about these decisions. My trusted agent, Anne Dewe, did what she could. It was on 19 March 1997 that she phoned to tell me that a paper-back copy of "The Byrom Collection" would soon be printed. That was almost one year after the 'spring' decisions of 1996. By 25 April my brilliant editor, Tony Colwell, was discussing with me the need for an up-dated Preface for the new edition. He needed the draft quickly, so on Monday 28 April I wrote in my diary:[1]

> I spent the afternoon preparing explanatory notes re: Preface and writing the accompanying letter to Tony C. By 4:30PM it was all ready. I phoned Cape. Tony was on the phone. Kirsty, secretary, agreed to deliver the six pages of Fax directly to Tony. I was relieved to send it on its way and to know that it is now at least on Tony's desk as promised, after he had requested it.

The very next day Tony was on the phone with me, his manner firm, assured and protective. The jottings in my diary capture his authoritative mood. I think it best to report it as written at the time:[2]

> Before 12:00 Tony Colwell phoned re: Preface (faxed yesterday). He thinks it's all too tame. The readers will not/cannot understand the subtext, and really he feels that since we are all having to go to so much trouble, it hardly seems worth it if I am not prepared to be more forthright about it all. I said I did not want to make it more difficult for Mark R. and that there were matters related to the archaeology which it would be premature to talk about. Although he understood my point of view, he still felt that I could find some way of making it all stronger. He said that I had been treated very badly and that everyone had been keen to 'put me down' – 'Hildy, Salford housewife' etc. I told him he was 'winding me up.' He was, however, keen for me to think again. I spent the next two to three hours weighing up the letter I sent to John Peter, journalist, The Sunday Times, in October 1995.

Elizabethan Theatres. The Parametrics

The models made on view

43

(Allan was in hospital at the time – it was handwritten.) I sent a copy of the book and a copy of an earlier letter to Theo Crosby. I decided to rewrite the Preface using the 'Peter' letter as a framework. The academics have not paid proper attention to the drawings and this has to be said. My quarrel with them is that they have not attempted to study the originals to see whether I have omitted a crucial factor. [They did not like the ground rules and were not prepared to play the game a la Hancox and the Trustees.] They have pretended the drawings do not exist. I am going to have to take them on, and they are now going to regret it. I will now have to go into training. It promises to be very exciting. And so the new Preface was faxed at 6.45PM!

The next weeks of May were peppered with phone calls from Tony, requiring my full attention to preparing illustrations and to other publication details. The missing book negative certainly created more work. And then there was the question of a Publisher's Contract for the paperback edition. Nothing had been mentioned to me about that until 10 May. After more deliberations with my agent and others I signed the document on 14 May. It was to be a new beginning.

By this time, I had accepted an invitation to the Official Opening of the Globe on 12 June for myself, Allan, and our son, Julian, my agent and her husband, Anne and Roddy Dewe, and my editor and his wife, Tony and Margaret Colwell. I included them in my party as my way of saying thank you for their support during what had been a difficult time. We had all earnestly hoped that the paperback edition would be available before the official opening.

That did not happen. I had not received the 'proof-copy' ready to read by the day before the Official Opening, Thursday 12th June. My diary says it all for Wednesday the 11th:[3]

> Tony Colwell phoned just after 6:00PM to tell me that there would be no point in me going on Friday. He had received a phone call from the printers to say that nothing would be ready for us to check. So much for all the work I have done with the illustrations. Actually, in one way I am very relieved because it would all have been such a rush and too much for us, especially Allan. So there we are! The hand of Fate has taken control. But what is the printer doing? The dates are getting pushed further and further on. The book was supposed to be ready and in the Globe bookshop tomorrow and we haven't even seen the PROOFS.

We spent the first part of the next day travelling to London for the big event, arriving at our hotel the 'Bonnington' early afternoon. We settled ourselves and welcomed our son who had travelled up from Bristol especially for the occasion. Tony joined us for afternoon tea, all of which occupied a couple of hours. My diary will take over from here:[4]

By 5:15PM we were up in our rooms preparing for our very special evening. I had a sort of dress rehearsal and Julian helped me to decide what to wear. I must say that his choice displayed a tastefulness that I had not realised he had! The menfolk helped each other with their dickey-bows and I took one or two photographs. We were getting into a taxi just before 6:00PM. As we approached Bankside the police presence and security became increasingly evident. Once we got out of the car the atmosphere was wonderful. The streets were lined with sightseers watching the rich and famous arrive! It was a very glitzy affair. One could see the aura of money around most of those present. We were frisked as we went in, ladies handbags only, men had to hand over keys and money. Allan was 'special' because of his pacemaker. Once in, the champagne was flowing. We all had glasses, but Julian and Allan had mine between them. We walked on to the patio, treading on named paving slabs as we moved around and then 'in.' Goodness! Julian and I had the best seats in the house – centre balcony. Allan and the rest, Anne, Roddy, Tony and Margaret – side, stage left. Soon the distant, eerie drumming began. It was wonderful. The stage was dramatic, although I didn't like the 'hangings.' The Queen/Prince Philip arrived by barge and sat to our right in the 'Gentlemens' Room.' Zoe Wanamaker opened the proceedings. It was a proud moment. A wonderful evening. I thought Mark's performance 'moving' for several reasons. I have made a cassette recording of the event… After the event there were fireworks on the Thames and then a champagne supper in the 'Undercroft.' I will never forget it. I do not think any of us will. Hildy and wife were there, unsuitably clad. They moved among the 'groundlings.' I had a word with McCurdy. He looked wan and dour. I met Peter Kent, a fund raiser, and talked him through my own story and connections. He listened, I know. He had borrowed the book from Anne. I chatted briefly to Claire and Lucy Beevor, and Tasha and I had a good bear hug, Memorable all of it. Roddy's chauffeur drove us back to the 'Bonnington.'

The next day was to be 'proof-reading day' at Cape with Tony. No proof-copy available. No meeting. So we spent some time at the Public Record Office in Islington checking up on facts regarding my research projects. The evening was a time for reflection with respect to the 'Opening' and preparing for Saturday's celebratory performance of Henry V at the Globe. My diary entry for Saturday reads:[5]

We got a taxi just after two. Looking forward very much to this historic occasion…we were there in good time. Today I took my camera with me and took a few photographs before the play started. We joined other revellers in the courtyard outside. There was plenty of champagne again today. We had a quick look at the new restaurant and noticed McCurdy and his wife, Jon Greenfield with a different group and other familiar faces. Soon it was time to go in. We had wonderful seats again, immediately next to the

Gentlemen's Rooms. There was a party from New Zealand to our right and a group from Toronto immediately behind us. There was great excitement all round. Henry V was memorable. Mark's performance extremely moving, delicate and spiritual. The production competent. Costumes – stylish and tasteful (apart from Mark's crown, too deep). We loved every minute. At the end a moment's silence for Sam Wanamaker, and a moving speech from Mark again! Michael Holden stood to his right, I suppose in attendance, for support, for what reason? At this moment I decided I could not go on to the Founders' Dinner in the 'Undercroft,' no matter that we had paid £70 and £70 for 2 places. It would have been dishonest to be there. I had seen Mark's production and I am happy with that. But, to have met the eyes of Jon Greenfield whilst eating a festive meal, across the table, would have been something that would have made me very upset. We walked away to London Bridge, got a taxi and had buffet fare at the Bonnington instead. Allan was very supportive. We loved the occasion in every other way. Hildy and wife were there – 2nd Gallery.

The next day our train journey home was relaxed and subdued; we were both very tired after our exciting few days in London. My diary entry summarises some of my thoughts for that day:[6]

It was indeed memorable in so many ways. I am glad to have seen the Queen and Prince Philip at such close quarters. Strange to have seen her without her usual trappings. Mark's performances were memorable – the fireworks, the champagne, the ladies' dresses, having Julian with us, the Undercroft. I will not forget either, having to make the agonising decision NOT to attend the banquet. I felt absolutely dreadful. But for me I know it was the right one to make. Those responsible for making decisions would have been there. It was too much for me. Come to think of it, I have seen no sign of Gurr and Orrell – plenty of mentions in the booklets. Perhaps they went to earlier performances? The cost of it all. It was an experience I wouldn't have missed, but what next?

The proof-reading day had now been fixed for Wednesday 18 June, less than a week after the Official Opening of the Globe. But the Globe was now 'up and running,' its success as an enterprise all but guaranteed. For me nothing had changed. My forthcoming book would be saying as much. For some people it would be clear – in my eyes the design of the new Globe as it stands is a 'near miss' and not accurate. Perhaps it was thought best that the book should not be available at the time of the celebrations. But who would have had the power to make such a decision? I set off for the meeting at Cape for the proof-reading with all the enthusiasm I could muster, carrying my precious working 'proof' illustrations immaculate and secure. I also carried my latest workings on the Bill McCann findings, along with his accurate illustration of them, sent directly to me this time.

We had kept in touch since our first meeting in November. He had already given me permission to use his work in conjunction with mine in any publicity surrounding my own theories and the forthcoming paperback edition. I had spoken to Dr. McCann on the phone at some length only the day before. My findings seemed to be in line with his latest radar imaging. The drama of the day is now taken from my diary:[7]

> We arrived at Cape at 1:15PM. Tony came down to collect me. After opening pleasantries, he quizzed me for any up-to-date news. I said that McCann had opened the flood gates yesterday re: illustrations. We were by this time in his office. I gave him the details of our telephone conversation. With steely eyes he asked me what I wanted to do about it. We talked over possible journalist interest. He muttered that they might not get it right for me and it wouldn't be properly understood. Did I want the drawings in the book? I almost visibly gasped and said that surely it was too late? He said "what would you want to put in?" I hurriedly thought and said two drawings, 1) McCann's and 2) mine, superimposed and then an explanatory note. "And then what would you leave out?" Gosh! What a question. I thought that Nigel Pennick's contribution could be managed without. It meant three pages 'in' and three pages 'out' – a bit of the astrology would have to go too. We sat there for a few minutes and mulled it over. I said we would have to have permission from McCann. There then followed a hair-raising series of telephone calls. McCann had to fax agreement not just to Cape but to me as well. I had to write a piece on McCann's behalf using his words from the Interim Report and putting in a little of my own, getting him to agree to it all on the phone. Tony spoke to him – it was all very professionally done but quite hair-raising. I could scarcely believe what had happened. We didn't begin dealing with the illustrations until 5:00PM. Allan came back from his Globe trip about 4:00PM. I was then whisked away to see Gail in the publicity department for about twenty minutes. A cool customer indeed, but efficient. Checking the illustrations was no easy matter. Allan was a big help. We were so short of time, and so he was extracting the illustrations from the 'Collection' quickly and to save time. Fortunately, I had shown him yesterday how I had filed the drawings. We were rushing away from the building at 6:10PM to get across London in time to catch the 7:00PM train. We had just enough time to grab a sandwich and a beaker of tea to have on the train. Phew! What a day, but very exciting and exhilarating. What a turn-up for the book – literally!!

I was by this time convinced that someone was 'marshalling the programme' surrounding the publication of my books interpreting the Byrom drawings. The forthcoming paperback would be another issue for worry because it contained new material directly related to the original site of the old Globe. My interpretation of the new evidence would become very relevant – a real conflict based on the physical findings. The conscientious intervention of my editor Tony saw to it that would be the case. The truth would be out. It became abundantly clear to me that he had nothing to do with the missing negative of the hardback edition, nor the selling of the stock in 1996.

Returning home, I thought it would be a question of waiting for the new paperback to appear. But not so! More delays by the printers, Butler and Tanner. The actual launch day was 7 August, seven weeks after that significant meeting with Tony on 18 June when he gave me the opportunity to include my work regarding Bill McCann's actual findings in the book.

During these intervening weeks, sensing my despair, Tony became a wonderful ringmaster, taking initiatives at his Cape office with aplomb. He has often been described as a scrupulous and exacting editor and I too can vouch for his style and exactitude. Whilst I was receiving more packages containing plans from Mark Rylance, all requiring immediate attention, he was dealing with careless errors by the printer and matters related to publicity. One such frustrating day is captured in my diary for Thursday, 26 June. Tony had attempted to speak to Dr. McCann about the caption to his illustration of findings:[8]

Bill McCann's Office – he was on leave. Tony C. used the opportunity to have a flurry of calls in connection with the printer and reviewers. There is now going to be another delay and would I please let him have a list of 'specialist' magazines etc. which should receive advance copies? I'll let him have it by tomorrow. At about 4:30PM George Murcell phoned. Suzie had been trying to send a Fax (7 times)! The content of the fax was a résumé of a lecture given to the Royal Society of Arts in June 1996 by Jon Greenfield, Mark Rylance and Michael Holden. [Topic: Globe]. Jon Greenfield's stance was based on proportionate geometry (Cabala) and Hermetic Symbolism. Oh dear! Oh dear! How could he? I do NOT recollect him being so supportive before! When he visited here in November 1995, he was 'anti.' It is as though the book and the truths within it are being used, but the measurements are not properly accurate [new Globe].

I wondered at the time how the two relevant illustrations, fresh additions to the paperback, would be received by this triumvirate. The days rumbled on uneasily. Exactly one week later I received another surprise package. It was a copy of a book entitled "The Arch and the Rainbow" published by Lewis Masonic Books. The author was the Revd. Neville Barker Cryer. It was a reference book in connection with the craft of Freemasonry and was presented in a very elegant format. When I looked to see who the printer was, it was Butler and Tanner. Quite suddenly, memories rushed back. In the 1992 issue 104 of the prestigious international Masonic research periodical 'Ars Quatuor Coronatorum' was included a review of "The

Byrom Collection." The article had been written by the same author, the Revd. Mr.Cryer. The printer of that journal was also Butler and Tanner. By that time, I had come to know the Revd. Neville Barker Cryer well. He had been a research consultant of mine for a number of years.

As already mentioned, the firm had also printed the 1992 hardback edition of "The Byrom Collection." Not long after its publication, the book had appeared as the centre piece in the window display of Toye, Kenning and Spencer, the Masonic book and regalia shop at Great Queen Street, London, located directly opposite the headquarters of Freemasonry. It seemed that Butler and Tanner were the printers of choice for quality books, including Masonic reference books. Carelessness, delays and missing material seemed totally uncharacteristic of this firm. It was agreed that a letter should be sent to them expressing concern for how the material now being prepared for publication was being handled.

Exactly one week later, on 10 July, I received my first copy of the paperback, one month after the official opening of the Globe. It was a quality production, as expected, though late. Almost one month after that I received my supply of books for distribution. Fortunately, just in time for the planned official launch of the book by Cape, Thursday, 7 August 1997. My diary captures the day:[9]

> It was really something of a non-event as an occasion. I got on with the business of the day. We posted the parcels I had wrapped up yesterday and went up to Prestwich early to post them. Later in the day I also posted books to Mark Rylance's mother (Anne Waters) and Marke Pawson. So, a lot of books went off I'm glad to say. The packing is a chore but never mind. It is a question of spreading the Gospel and there is simply no other way of doing it at present. We had run out of bubble paper and so Allan collected another roll in the afternoon when he went to collect copies from Waterstones and Dillons. Dillons hadn't got any [of my books], none on order so they said. At Waterstones the book is in the architecture department, buried under others. I rang Stewart Wilkie, the Cape rep., and talked at length about the poor publicity locally. I mentioned the programme last evening (One Foot in the Past – Byrom's birthplace etc.). I think that he was suitably embarrassed and has promised to get publicity cards etc. made. But of course this all now after the launch day, including the new Globe. Painful. Painful. Painful. I had no callers from 'News Desks.' I made several telephone calls – points of interest. Lottery to be filmed from the Globe, London on Saturday. Goodness me! What implications does that have for me? Good? Bad?

> Thus far there had been no reviews or publicity that I could associate with the paperback or my project in general. I watched the Lottery programme on Saturday, 9 August, two days later with a degree of resignation.

So the enigma of the book carries on. I am certain that there will be no reviews tomorrow either. For the moment the Globe issue is a dead issue (wrong design). I watched the programme with horror. Michael Holden talking to the Presenter (Ann Gregg) about the concept and a twelve and a half million pounds lottery grant. Patrick Spottiswood, the education officer, plugging that aspect (education). A male and female actor and actress chatted through briefly how 'big' it all was. I really am in a dilemma (conscience). How should I proceed now?

As usual, my diary records my reaction to the programme. I had to face the inevitable. Cape had fulfilled its obligation in producing the paperback. The Globe administration had completed their task of building a theatre dedicated to the legacy of Shakespeare for the world stage, fulfilling Wanamaker's dream, using in part public funding.

I could certainly understand how those responsible for the Globe undertaking were determined to promote its success. Anything that got in the way of that mission would need to be minimized, and so it was. By now, the publicity machine had swung into action. Daily, there were reports on the televised media and in the press. By Friday, 15 August, during a telephone call to Tony, he expressed his real concern about the 'silence' surrounding the paperback. His experience in the publishing world over so many years was signalling to him that the marketing of the Byrom project was not receiving much interest. But this was not just *any* project. I viewed the Bankside Enterprise as a 'masterpiece' of accumulated ambitions gratified, regardless of whether the 'anchor' of it all was based on the core truth of the design of the building. It was clear that those who were responsible for the decisions made did not understand that the original building reflected 'eternal truth' in its design, which goes to the heart of the authentic Shakepearean experience. To me it was a sophisticated 'Folly.'

I recalled meeting Theo Crosby for the first time in his Pentagram office on 12 June 1989, and his enthusiastic response, together with that of his colleagues to the proposed new Globe as I interpreted it. 'It' was officially opened in the presence of the Queen and Prince Philip, exactly eight years later to the day. There had been plenty of time between those dates for serious studies and deliberations to take place by those designated to do so for the Wanamaker Project. Even though Theo died in 1994, I had gathered supporters around me over those years, scholars and enthusiasts, as well as having offers from media companies for documentaries. All had read my book(s). Some had seen the Collection in person. But there was only one Globe in London to be made faithful to 1599. Patterns and plans *did* exist. My researches were undertaken with that firmly in mind. For me, it was not a matter of economics and I had no Patron. I had a sincere obligation carried out to that end and I was prepared to be a team player.

Mark Rylance, something of a free spirit himself, made a stubborn heartfelt effort for my inclusion and perhaps was responsible for a small order of the paperback to be placed in the Globe bookshop when the paperback came out. I thought of it as a goodwill gesture, considering the obvious prejudice that had been displayed towards my efforts. I have always considered it a personal privilege to have had the opportunity to work with the Collection thus far and to have had a measure of control over the use of it through the licence agreement signed by the Collection's Trustees. Nevertheless, I also regarded the involvement that I had at several meetings and conferences along the way as something of a 'charade.' Who were the puppet masters and what was the problem? I would carry on. The integrity and solemnity of the knowledge behind the collection demanded that I should do so.

And so it was that on Thursday, 21 August, 1997 I had another special meeting with Dr. McCann. It was, by arrangement, to be with a curator of the Science Museum, London, at Olympia, in a separate building but part of the Museum. My diary entry describes the day:[10]

> Bill McCann was prompt, 8:45AM. We 'collected' ourselves, chatting in the lounge for a few minutes (Bonnington) and left for Olympia. The journey took longer on the Underground than I expected and after signing in we barely waited a few minutes before Kevin Johnson appeared. I vaguely recognised him. The [display] room was almost as large as my school's sports hall. No one else was around. We had been given white gloves for handling items. Johnson set out the boxes on a small table for viewing. It was extremely nerve wracking. My eyes searched for familiar markings on the brass plates. I got out my collection of acetates of the theatre drawings and began to link the Byrom copies with the brass identical ones. The Globe's "upper pt" is identical, brilliant, "Stars Mall," reverse matched another. And so it went on. I thought that there was a parametric 'Globe', not quite, but parametrics nevertheless. I got out my '72' measure and demonstrated that the measures were identical. Were there any more? After searching the computer, Johnson was able to tell us that two were in the Science Museum itself on display, second floor, Maths Department. He photostated copies of all the plates – dreadful repros, but they will be of some help to me – exact size and enough for me to recognise them. I also took some photographs, having recognised how poor the photostats were. Thank goodness I had film. I asked for permission first, of course. Bill McCann was stunned. "A portcullis has dropped down in front of you."

The immediacy of Bill's response, one of complete surprise, I took note of. He recognised the accuracy of the 'matching' between the Byrom theatre drawings and the brass plates. The exercise was also witnessed by Kevin Johnson, a member of the staff of the Science Museum. It was

a unique moment of identification to all of us and was new information
to the Museum. There had been a time when the general public had
been given the opportunity to submit explanations as to what the brass
plates represented; they were certainly rarefied in their geometric detail
and obscure in themselves. It was the Byrom drawings that began to
give them an identity. It was the Byrom Collection that would go a long
way to helping unlock the mystery surrounding their provenance. But the
drawings were on paper and card and unlike the brass plates were more
easily destructible. Apart from any other reason my sense of responsibility
moved up a notch on that day!

Dr. McCann was a distinguished professional in his own right.
Working for the Museum of London, he supervised and managed
large archaeological projects and established and managed The Clark
Laboratory for the Museum, focussing on Ground Penetrating Radar.
Having committed myself to including his latest findings of the old Globe
in the paperback, I had earlier arranged a meeting with him and my editor
at Cape on Friday, 25 July, before Launch Day. The spirit of the day was
recorded as follows:[11]

> Having had lunch, we were sitting in the lounge of Cape at 1:40PM and
> at 1:45PM we were in Tony's office. It was nice to see him. We exchanged
> 'beams' and then he presented me with the illustrations and galley proof,
> all very efficient. After about 5 mins. Bill McCann arrived. We didn't hang
> about – straight down to the Board Room. Very formally, I announced that
> I was going to present Bill with a special 'Globe' selection of acetates. I set
> them out on the table first and then demonstrated the Globe's 'upper pt'
> and parametric 'display' of complete compatibility, and then the rest of the
> sequence in like manner. I had said that I did not expect McCann to express
> an opinion, but then he did. I was inwardly delighted to hear him say in front
> of Tony that he fully recognised the 'indisputable' evidence of the Globe
> drawings. Very quickly he said we should do a paper and some articles for
> reputable magazines. He was so excited. I kept the whole exercise formal,
> thanked Tony/Cape etc. We then went down to the publicity department to
> meet Kate. Tony had told her of Bill McCann's excitement already and so
> to a lesser degree I demonstrated the basic two (para and 'upper') again. We
> also discussed some aspects of the 'suppressing' side of the archaeological
> work, and the drawings. She is an intelligent young woman and wondered
> whether an investigative journalist should take the story on. She thought it
> should be brought out into the open. We left Cape just after 4:00PM. The
> meetings had been really worthwhile. We just had time to grab a cup of tea
> at a nearby café and then made our way to Euston.

It was after this meeting and Dr. McCann's reassurance regarding my
Globe design measurements that I made the conscious decision to introduce

him to the brass plates in the Science Museum. In 1990 I had learned of their existence, but given the apparent lethargy from scholars in general, I had little inclination to waste more energy on what appeared to be of little real interest to them. Clearly, I needed to confirm the compatibility of the Plates and the Byrom drawings in detail. That task was undertaken and completed on 21 August 1997 shortly after the meeting. I was lucky to have had his assistance in preparing for the launch of the paperback edition of "The Byrom Collection" and I will be forever grateful for his moral support.

Much effort from many people from very far and wide over many years had gone into the planning, construction and presentation of the new Globe enterprise. The original ambitions of Sam Wanamaker were wholesome and genuine. Despite the many difficulties he had to overcome, he was supported by his faithful scholars, experts and volunteers. And then there were the Patrons. Everyone knows that he and his loyal architect, Theo Crosby, died before the dream was realised. But the new Globe stands erect as a monument to human endeavour whatever the cost.

At this time, for a while, I felt guilty that the brass plates had emerged as a seal to what I believed were the blueprints and formulas of a group of temple-like buildings in the City of London built in the sixteenth and seventeenth centuries. I felt like a fly in the ointment. The sequence of drawings and plates included the Globe and its predecessor, The Theatre (1576). My published books had already made that clear. I knew that a portcullis had been dropped down in front of me.

Some thought I was mistaken, but there was a growing number of people getting in touch with me, all intelligent, mostly worthy and dedicated researchers who firmly believed that my interpretation was correct, backing up the Claus Visscher engraving of Bankside with its panorama of playhouses. Those people included Professor John Gleason, the theatre owner George Murcell and the author Iain Mackintosh, who published "Architecture, Actor and Audience" in 1993. As a theatre historian and design director of 'Theatre Project Consultants,' Mackintosh chose to include a description of the concept behind the Byrom theatre drawings in his book discussing theatre design concepts through the centuries. At the time when he was preparing the text and then publishing it, I was in continuous contact with Theo Crosby. Mackintosh refers to "The Byrom Collection" in his book. I have never met Mr. Mackintosh or spoken to him, but I had to consider his contribution, along with many others at this critical juncture. My diary for 28 August, a few weeks after the meeting with Bill McCann and Tony in 1997, captures my opinion and my decision at that time:[12]

And now my task begins! The case for the 'Globe' (design) is proven beyond a shadow of a doubt, but how do I bring about such a disclosure? My book is out. But that will not be enough. I need to make a totally coherent case for the manuscript "471" [associated with the brass plates] and Science Museum collection, along with the archaeology. It is an enormous task.

MS 471, I had been told, is a small mathematical text considered to be a part of the brass plates phenomenon. However, my task must centre round the brass plates themselves. My '72' measure verified the exactness and precision of the geometric structures present in the plates and the drawings. So who was responsible for them and where were they made? The Science Museum could not tell me. I would have to look for my own answers and I could not rest. I needed to set the record straight.

I had been surprised at the lack of reliable information available about the brass plates and MS 471 in the Science Museum. But then on reflection, perhaps I should not have been that surprised. They were from the same tradition as a section of the Byrom collection and the mystery surrounding them had taken years to resolve. The Collection itself was thought at one time to have been owned by the British scientist Robert Boyle. Boyle died in 1681.

I was not going to delay my quest. Two days after that decision, on Saturday 30 August, my diary records my subsequent efforts to learn more: [12]

Today has very much been a working day, with much of it being spent at Manchester Central Library. We got there at about 10:45AM. The latest revelations from the Science Museum necessitated that I should attempt to learn more about Wenzel Jamnitzer, the goldsmith, whose book of 1568 figures prominently with the plates. I also wanted to study Boyle's book on 'perspective' published in 1735. Manchester has a wonderful collection of books and manuscripts, for I was able to look at both. But before looking up anything, I presented the Local Studies Department a copy of the 'paperback' and a few notices announcing the publication. Perhaps they will pin one up on their notice boards, we will see. I was able to make lots of notes and Allan photostated some of the most important pages on Jamnitzer. He was obviously a very distinguished goldsmith with workshops in Nuremburg, serving some of the most important European families. One of his sons was killed in the riots of 1572 (Huguenot). I must find the link of how the book came to England, and by whom. I suspect that we are back with John Dee and Bartholomew Close and of course, the Leicester/Dudley family.

Wenzel Jamnitzer died in 1585. His son Hans and grandson Christof Jamnitzer continued his business. John Dee visited Nuremburg in 1587, having by then spent time as a tutor to the Leicester/Dudley family. I was already certain that the Byrom theatre drawings were closely associated with the de Bry Huguenot family. The day had been well spent. Reflecting, I felt energised, but that energy was dissipated the very next day by the tragedy of Princess Diana's death.

CHAPTER SIX

Disquieting Uncertainties

PERHAPS IT IS as well that we cannot always be prepared for the unexpected. The element of surprise for good and bad is always present. These last few days had been heavy with grief for much of the country. I could not have been prepared either for what came knocking at my door after 31 August. It was the 'postman' on Wednesday 3 September. My diary extract records the telling tale:[1]

> I had such a shock, I want to say surprise, but it was a shock, when I opened a package in this morning's post. There were twenty-one pages (I counted them later) from Mr. Barrow man himself. I had not heard from him since April 1993! [The Meeting]. I didn't read any of the pages. Allan glanced through. Pages of verses and lots of additional signs and drawings. I felt sick! The date on the envelope was the 30th June. The package had been sent to Cape. Around that date I was having trouble over the 'paperback,' delays and mess-ups, Butler and Tanner. See the diary entries around that date. This seems to be more than a coincidence. There is a connection somewhere. WHO is it that has chosen NOW to write BEFORE the 'paperback' publication and WHY? Very mysterious.

The Launch of the 'paperback' had been on 7 August. The package had been posted before on 30 June, shortly after the opening of the Globe Theatre on 12 June. It had been addressed to me care of Cape. They sent it on to my agent. The package took nine weeks to reach me.

The accompanying letter was signed by the same person I met on 6 April, 1993 Brian Charles. There had been no contact between us since that date and the arrival of the package on 3 September 1997. I had decided in 1993 that he was a willing 'postman' for someone else unknown to me. With that very much in mind I had not followed up on the suggestion that we should meet again. But here, over four years later, posted from the same address, '3G, Devonshire Buildings, Barrow,' is the messenger again.

The contents were immediately troubling. It was not a question of reading through a document of twenty-one pages. Each page was a topic complete in itself, sometimes with information on both sides. Some pages included illustrations, with phrases in several different languages and signs representing various cultures. One could sense an unusual intelligence, but there was no way that any of the pages could be described as straightforward.

I spent some time over the next hours striving to recognise some sort of theme or 'pulse' to the collection of pages. As the hours and days ticked by, I came to regard the contents as menacing, secretive and written in code. I did not feel like talking to anyone about this latest communication immediately. However, I did ring my agent to tell her of my concern at having received such correspondence, which had been re-directed from her office. Unnerved by thoughts of having to deal with the interpretation of such material, I put the papers to one side on that day before the funeral of Princess Diana, Saturday 6 September. I allowed myself Sunday to deal with family matters. Monday, the 8th was recorded as follows:[2]

> I have worked on the 'Barrow' correspondence for most of the day. As the day progressed I became more certain that whoever is responsible is determined to let me know of their very real concerns, understanding and involvement. They know me, what makes me tick and my sadnesses. How do they know this? I can only conclude that I have been watched and studied and recorded for a very long time. I feel, just now, very unnerved. I will have to continue working on the documents because I cannot persuade myself, convince myself, with certainty of the intentions. It has to be the telephone in the first instance, but there seems to be a deeper understanding of me and my personal experiences. [Regarding my Research Project.] I already live a recluse-like existence, as this diary will demonstrate. What is to become of me? What am I supposed to do? I phoned Tony Colwell at about 11:30AM and told him of the package. At first, he was rather dismissive. I insisted on making him listen to the dates, records. He admits to finding it curious but doesn't think it at all helpful/pleasant. I wonder whether he really does understand? He has been very shrewd recently. I think that I may have convinced him, but only eventually. It is so draining.

My conversation with Tony had not really helped, because everything to do with the package could seem to be bogus to an editor like him when talked about over the telephone. Tony's reaction served as a warning to me. I would need to be discreet. And so, on to the next day, Tuesday the 9th. My records show how I felt:[3]

> I was so very tired this morning, emotionally and intellectually drained. This is my personal diary, so I feel able to, and must commit my very private thoughts to it, on this occasion, as they unfold. The realisation of what the contents of the 'Barrow' envelope contains could be frightening. I am not frightened except it is at this stage very unreal and disconcerting, when one learns that there is a 'third eye' somewhere recording every detail of

my private and personal life. Of this I am now certain. What the aim is, or why it should be happening, is not clear. I understand now that all matters related to the book are being 'controlled' and I am being 'contained.' I have spent all day working on the transcriptions and they become more illuminating by the hour. Thank God for the sanctuary of Number 11!

It was page nineteen of the twenty-one pages that first gripped my attention. So far into the mass of information connected with the earlier pages, why did it take so many hours of painstaking attention to recognise the basic theme of the package? I present a photograph of the page here in an effort to help the reader achieve an understanding of the predicament I was facing at the time. I also include a summary of the same page as I understood it.

But first, the title 'A Truth Lies Standing' represents the newly built Globe theatre. The implication being that it does not represent 'a truth' but stands as a lie. The 'piece' is directed to me and my understanding of the design concept. The author of the 'pieces' in this package created them with the idea that I would be able to fathom what he intended me to understand. He would know that I had written and published a biography of John Byrom ("The Queen's Chameleon"), based on an original manuscript of Byrom in his own shorthand, not published in his day. I taught the shorthand to myself using an instruction book published after his death in 1763. Using the same principles of decoding or perhaps 'coding' in the author's method, my understanding of 'A Truth Lies Standing' suggests the following as my interpretation:

SUMMARY: There are those who think my understanding of the core design concept of the original Globe theatre is rubbish and that view spread quickly. There are others who think differently, and my destruction cannot be merely a planned event by active academics with unpleasant intentions. There is something bigger here. The author of these 'pieces' has studied Archimedes' theories with the intention of understanding Byrom's geometric papers and my interpretation of them. He's telling me that my own interpretation of the parametric three-dimensional drawing was a triumph in my favour. That drawing is the key to the Globe sequence, as I see it. But there is a need to conceal certain aspects of the philosophy behind this drawing (and the other drawings), known it seems, to some very scholarly Freemasons. Cryer's meetings with me, which he seems to know about, will have provided clues, he thinks.

The verses from line 14 onward in this poem can be understood more easily. The author is implying that I should aim to talk/tell of this precious treasure and the special knowledge contained within it, but unless I understand the technique of presenting it in an 'acceptable' way, I will

A TRUTH LIES standing

Mute cadent pile {sum/some} do say U say
steep in mystery NEWS gabble every way

Emir/immer} of the heavans Basálica IV·square
trivial facile truth raised up & posted in the heir

your ruin cannot be compassed by Muses with horns
circumspect the son dis played im paled with crown of þ

a cloud of stone seen & soon forgot by idle tour ist aye
OR geo metric mounting place for Jacobs staff & eye

Labor I Tri the art & handiwork of ArchMedes ape X dialis
Tri umph of the intellect over mundane earth altar/alter Tri alis

Maçcon the bones of reality mus hide from human eye
in eryre lime stone cave Dis played the say that Jimply

So look behind the hard façade disclose eternal treasure
talk is cheap & so is verse unless you know The measure

truth & falsehood inter twined & in here J do esconce
the art is how to weigh & weight each word & speak them both at once!

if J say J tell a lie most people do believe me
if J say J tell the truth you think J would deceive ye?

If Mochahamed will not attend the Mountain of the Magii shown
then the mock mountain must needs attend κοινοδικειν τη η ΚΛΙΡΟΣ town

A Truth Lies standing

not succeed. The author thinks he knows the answer and wishes to work with me. He then suggests that if 'Mock-a-maid' (me) will not meet the 'Magii' at the mountain (Hill of Hoad, Ulverston), then I should arrange to 'attend' Keith Critchlow of the Kairos Foundation. This foundation studies the 'Quadrivium' of arithmetic, geometry, music and astronomy.

I knew of course who Keith Critchlow was. I had referred to one of his papers on geometry in "The Byrom Collection." But what was significant, on receiving this package and seeing the reference on page 19 to 'Critchlow,' was that I could not help recalling a very important commission immediately prior to the book launch at Groucho's of the hardback edition on Thursday, 14 May 1992 (5 years earlier). It began on Wednesday, 6 May with a phone call. It was a request for two copies of the book, one for Prince Charles. My diary records the details:[4]

> About 4:00PM Nicola rang [Cape publicity secretary]. Keith Critchlow had phoned, accepting 'Launch' invitation and requesting a signed copy to be delivered to him as soon as possible. She did not know who Critchlow was. When I explained that he was the Director of Research of the new 'Architects Institute' (Prince Charles), it sent her into a tizzy. She said that she would telephone tomorrow with suggested wording. [Thursday 7 May.] About 12:40PM there was a call from Cape. I have to put in an inscription [on one of the two copies].

> > To Your Royal Highness,
> > The Prince of Wales – with the greatest respect.
> > Joy Hancox – 8th May, 1992

> They would like the copy delivered (almost) immediately and so Allan will take the signed copy tomorrow direct to Cape, who will then arrange for a messenger to take it across London to Critchlow who will re-direct to Buck House. The second copy is for the Institute's library.

The Groucho's reception was such a busy affair I do not recollect being introduced to Keith Critchlow. It was Cape that had sent him the invitation. I never heard from him directly. Now, five years later, the Barrow messenger is making a bold and informed suggestion. Beginning to sense that the latest correspondence was deliberately timed to reach me on 3 September, Barrow seemed to be far away and simply providing a postal address to deflect from the mission of intent. Brian Charles was the trusted 'postman' and the designated messenger, I conflating them both. I continued to look for further clues. Who was there for me to complain to? Most of the correspondence was so enigmatic at this stage that no official body would take any of it seriously and perhaps suggest that it is a nuisance only, no more – ignore it. I could not.

It was page 13 (in addition to page 19) that provided me with the kind of information that I could well have managed without and caused me to confirm that all was not well. My years of contemplative study, consultations, effort and expense were being silenced. It intimated that the purpose was so that the perceived view of London's new Globe would no longer be challenged. I include page 13 here as an illustration, followed by my observations of the contents of this disturbing message.

I now recognised that my correspondent, the author, was using language in a variety of extremely clever ways. I had to be prepared to look up and trace definitions and meanings that were not immediately familiar. Sometimes he played with words and phrases, such as in the case with page 13. The title 'Rev on nous mute ons' is an example. It has echoes of a well-known French saying 'Revenons a nos moutons', which when translated means 'Let us revert to the subject.' But with the words of the title split as they are, they could be interpreted differently. 'Rev on nous' when read, sounds like 'Revenue'; mute = silent; ONS = Office of National Statistics? And then it could also mean a 'Reverend' talking to his 'moutons' (mute ons) = sheep (flock). Of course, it could mean all the differing interpretations put together.

Byrom was very clever with his own shorthand, which was used as a secret language in his day with the great and the good. If challenged, he was able to say that the phrase did not mean one thing, it meant another. Once I had noted and accepted the system that my correspondent was using, it became a painstaking study of every letter, its position on the line, the size of the print, as well as the extra signs and diagrams used. Page 13 was informing me of so much.

For various reasons, my interpretation of the overall page was that the contents were to do with the handling and publishing of my book (gnosis). The printed two verses on the page provided me with a straightforward understanding of what I needed to address. For clarity I place them here for easier reading:

Put another password in,
Bomb it out and try again,
Try to get past logging in,
We're hacking, hacking, hacking.

Try his first wife's maiden name,
This is more than just a game,
It's real fun, but just the same
It's hacking, hacking, hacking!

Rev on nous mute ons
or: Cleopatrés gnosis differently shipped.

Tru paschal man while making certain conse inspections
made a coup to hold infinite curly ques ᷑
in ordered definite directions

setting his teeth into masses of woholy thinking jumpers
fix ED in his own brassy style ... in teg rated ᷑
cir cuit toi as √∞ numbers.

"Put another password in,
Log it out and try again
Try to get past logging in,
We're hacking, hacking, hacking

Try his first wife's maiden name,
This is more than just a game,
It's real fun, but just the same,
It's hacking, hacking, hacking" ...

Reals within reels
Wheals ——·"— wheels
deals ——·— Dials ...

Furnace Abbe △   *P.T.O.*

Derek de Solla Price's
reconstruction of a
calendar computer built
by the Greeks in
c.65 B and found by
sponge-divers off the
island of Antikythera.

Rev on nous mute ons

The commissioning editor of my books at Cape was Tom Maschler as noted. His first wife's maiden name was Fay Coventry. Was I, or had I been 'sent to Coventry?' Possibly. Being 'sent to Coventry' means being treated as a non-person, completely invisible, inaudible. I thought about it. Who was this person? It was extremely disconcerting and disappointing and I have since come to an understanding of it.

Immediately to the right and underneath the verses is a note of advice to the effect that business deals were being negotiated and recorded without my involvement and knowledge. At the time I knew nothing about it and these messages were pointing me in that direction. The name 'Pascal' figures strongly in the piece, he acting as the negotiator. Who was this character? Does the name refer to the Frenchman Blaise Pascal who was born the year of the publication of Shakespeare's plays in the First Folio of 1623? He was a man who suffered much from ill health, but published a book called "Pensees" (Thoughts). Pascal shows the greatness, and at the same time wretchedness of a man torn between his vanity and the humility of his faith. But Pascal is a character of the past. Who does he represent in the present?

Then there is another well remembered film producer bearing the name of Gabriel Pascal (1894 – 1954). He adapted the George Bernard Shaw plays 'Pygmalion' (My Fair Lady) and 'Caesar and Cleopatra' into films. Since 'CleoPatres' (Cleopatra) is used in the subtitle of this piece, this 'Pascal' has to be considered as well.

But then Pascal was the Christian name of one of the editorial staff who assisted Tony Colwell. He was always conscientious and pleasant. It left me with much food for thought for I knew absolutely nothing of the implications suggested here. At the time I signed my contract with Cape in 1990 I agreed that I would give no interviews until the time of publication (1992). This had been drawn up by Tom Maschler himself. My connections to and correspondence with Theo Crosby and Sam Wanamaker, which began in 1989, were not a consideration.

I turn next to the geometric structure being studied intently by the solitary Rodin-like figure known as 'The Thinker.' He seems to dominate the page. The caption at the bottom sent me looking to find out more about the Derek de Solla Price referred to. This caption is included in the text here for clarity:

Derek de Solla Price's reconstruction
of a calendar computer built by the
Greeks in 65 BC and found by sponge
divers off the islands of Antikythera.

I tracked the drawing of what appears to be some sort of machine to the Antikythera computer of antediluvian origin, thought to be part of a mechanism destroyed and buried in The Flood. As shown on the page 13 illustration, it was found by sponge divers and is believed to have been built by the Greeks c. 65 BC. The remains of the Antikythera mechanism are stored at the National Archaeological Museum in Athens. It is described as a complex clockwork with gears designed to follow the movement of the moon and the sun through the zodiac. Why my correspondent thought I needed to know about it at this stage was beyond me, but I continued to follow the clues relentlessly. There was a more serious intent than my cynical attitude registered. So to continue, studying the all but central statement alongside the rest of the page:

> Reals within reels
> Wheals * wheels
> deals * Dials

I was being encouraged to recognise and understand that from the perspective of my correspondent, a mechanism was in place for transactions to be arranged, negotiated and concluded regarding my work and its relation to the world at large. Apparently, this reference was intended to imply that a 'road-block' was in the works for the spread of the knowledge contained in the manuscript of my book. It was to be sent elsewhere for discussion, debate and gain. My correspondent felt that those involved were not equipped to handle such ancient and sophisticated knowledge. The writer signs off as 'Furnace Abbe' and leaves the instruction "PTO" next to that signature to advise me to turn over the page. Before doing so I glanced through the other pages and noted that my correspondent sometimes signed himself as 'Furness Abbe,' linking himself with the religious establishment, Furness Abbey. In this piece, 'Furnace Abbe' is his nom de plume, implying that he has become changed, incensed and hot/cross by the actions that were taken. To complete my interpretation of the first page I discuss the second as a matter of necessity and observation.

And so on to the next page: 'Furnace Abbe' heads it with a title, which interpreted reads as 'The Advisory Centre in Education, a course to study Middle Eastern Deities for ignorant students/individuals.' He goes on to say that he has 'launched' a new/fresh approach for those desiring to understand this information connected with mathematics, physics and astronomy at an advanced encryption level. At the time, this group of individuals was limited through lack of knowledge as to what was acceptable and were being influenced by a member who does not know better. (He is possibly referring to the decision makers of the Globe project.)

He then asks the rhetorical question: despite the 'chaos' in the variety of measurements being proposed, who was it that finally decided on the measurements of the Globe building without considering the Arabic numbering system as interpreted in the Byrom group of Globe drawings? As it turns out, that final decision has meant that all interested groups in society are being told that the agreed decisions are correct, and in doing so, influencing all matters to do with Shakespeare and what Shakespeare had to say to this 'madding' world.

* * *

Having selected these two pages from the twenty-one, I read these pages over and over as written and then my transcriptions of them. The writer knew what he wanted to say and recognised that I would doubt him. He included a few enigmatic facts which when understood would persuade me of his truthfulness and his closeness to the centre of where it was all happening.

There is a strong sense of a need for anonymity as well. He suggests that he would be able to help me make it otherwise. Who was this person?

Looking for any sort of clue, I spent the next few days studying the rest of the package pages and generally dealing with research issues connected with my books. I made up my diaries and wrote project letters. There were phone calls to and from Bill McCann, Mark Rylance and others. But as my daily diary shows, the Barrow correspondent became like a ghostly presence, with me constantly, all else temporarily paling into insignificance.

I managed to maintain my poise until the following Wednesday, 17 September, two weeks to the day after that last special delivery on the 3rd. That now-described correspondence showed that the years between 1993 and 1997 had been closely monitored by him. I finally decided that I would telephone Brian Charles's contact number with the hope that I could speak to him and learn something more directly from him. It was 8:30PM, not too late. My diary entry records what followed:[5]

Without any hesitation Mr. Harrison was quite happy to fetch Brian Charles to the phone. It was all quite spontaneously done. I introduced myself. He was completely thrown off his guard, obviously delighted. I first of all apologised for not having responded to his package earlier, but said that I had only had it for two weeks. I had spent a considerable amount of time studying the contents and did he want to see me about something? I gave him no time to think. His first response was he was very glad to hear from me. He thought I had disappeared from the face of the earth. Strange. Did he want to talk to me about the poems?

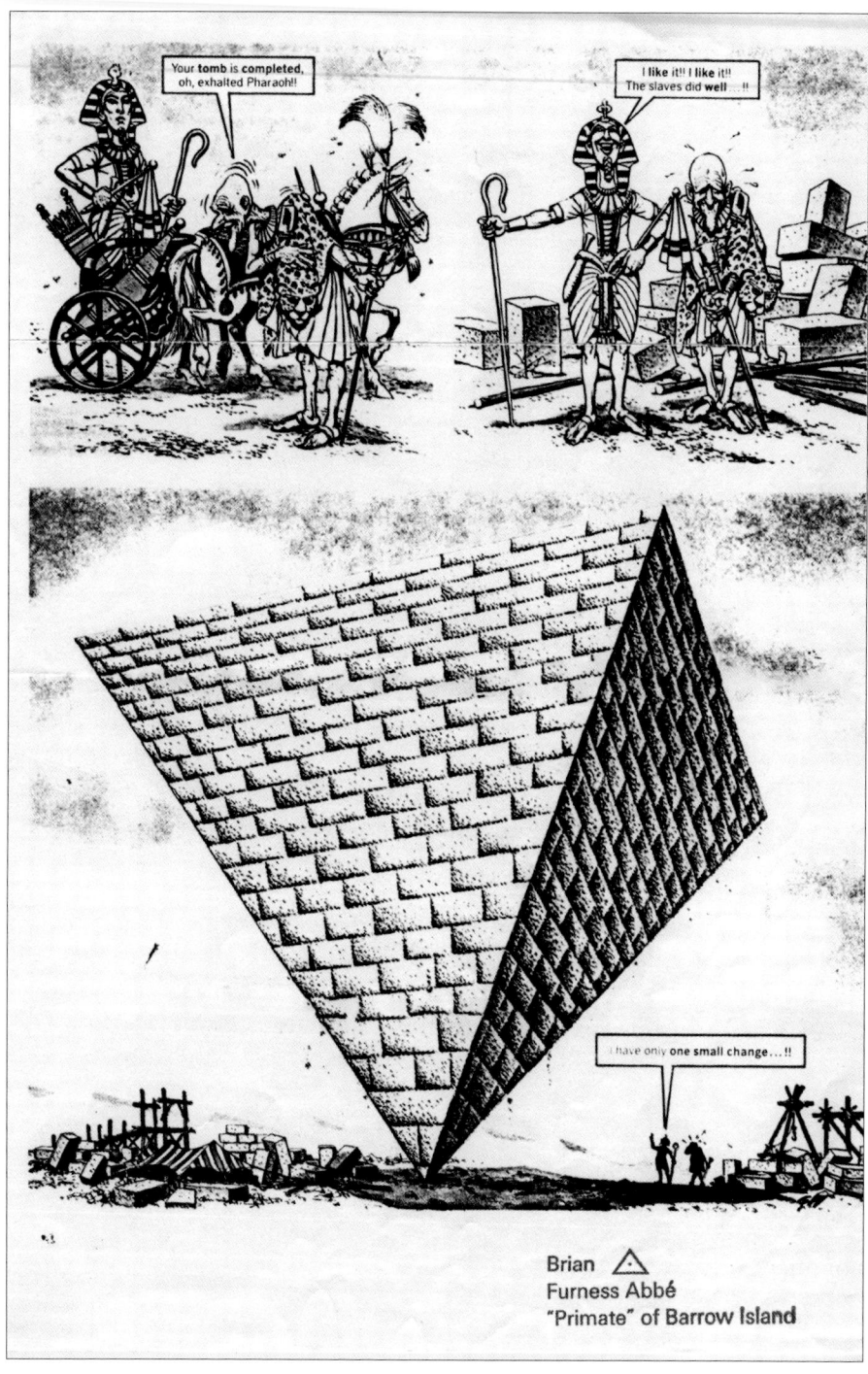

Your tomb is completed, oh! Exalted Pharoah

He told me that he was officiating at a poetry evening towards the end of the month; would I like to be there? I told him that would not be possible. He ventured that he had been ill six months earlier, but his health was now restored. I got no further with Mr. Charles, this 'postman/messenger' that evening. I promised to send a copy of "The Queen's Chameleon" for the contents of his package as thanks.

I decided that this fellow knew far more than he was prepared to say on the phone. Feeling uneasy about all matters surrounding the publication of the paperback and this recent correspondence, I decided to obtain a P.O. address. I did not want mail addressed to me lying around in offices waiting to be forwarded. In any case, I had no wish either for secretaries or others to be responsible for forwarding such mail. I felt as though I was involved in a very unsatisfactory situation so far as the Barrow correspondent was concerned. Taking control of my own mailing address would at least provide a safe line of communication without it being my private address. On Tuesday, 23 September, I was able to fulfil my promise and send a copy of the book.

Quite unexpectedly, on 29 September, only six days later, probably recognising my unease and the demands of protocol in my responses, my Barrow correspondent sent what was to be a final 'visitation.' It came as before via the post. It comprised one parcel and two separate letters; they arrived together. The parcel contained a block of oak wood about nine inches square with my name carved into it. Could it have been a sample from the debris of the Globe building site? Or, could it be (she says wryly) in his impatience he was likening me to a block of wood?

One envelope contained another group of five separate pieces. The tone and content, though rooted in the same topic, were showing an irritation at my lack of cooperation. The second envelope held one sheet only, a copy of which I include as an illustration here. It is clearly an allegorical representation of an Egyptian pyramid with crude satirical touches. As far as I was concerned, it represented the geometry of the Globe structure upside down. Captions state that the workers had worked so hard, but those ultimately responsible see a need for 'one small change.' I do not think I need to comment on other features in the drawing. I was frustrated to receive such a collective barrage of information and I consigned the block of wood to the refuse bin. I sent a formal acknowledgement to say that I would be away for the next few months and unavailable for further comment.

I received no more mail from Mr. Barrow. Nevertheless, I had registered the seriousness of intent which covered a period of more than five years. So I carefully mounted and preserved every piece of correspondence, including the envelopes. I also filed all reference material and research connected with it. It became part of my on-going archive.

The Hill of Hoad Geometry 100ft

The individual directing this 'messenger,' Brian Charles, was still avoiding detection and I regarded that as a serious matter. His knowledge of events and the proceedings that he chose to inform me about demonstrated an intimate awareness of core issues. He was also very much a part of the London scene, otherwise how could he know the maiden name of the first wife of my commissioning editor? Such a detail would not in my opinion be known to Mr. Charles from Barrow, with all due respect, unless he was personally connected somehow.

Despite the active silence that developed over the next weeks and months, an anxiety about my mysterious observer did not leave me. Still, I needed to turn to positive developments which were emerging. I was determined to do that. I came to learn much later that Brian Charles had died three months after my conversation with him on the phone in September. Records show that he committed suicide on or about 28 December 1997.

CHAPTER SEVEN

Onwards

I HAD ALWAYS hoped that truth and justice would be seen to support my thesis one day. My quest had become one of silent persistence. The Brass Plates confirmed the rightness of my work, to me, but the 'executive messenger' himself remained elusive. His willing 'postman' was now dead, yet I knew that this 'Messenger' (himself a messenger) was still there trailing my activities, for every now and again a telling coincidence would appear in the media or be drawn to my attention in an unexpected way. I called them coincidences at the time, but of course I knew they were not.

He was close by in some way. I knew not how it could have been possible, so I took no official steps to seek help to identify him. The years slipped by. I had by now focused my investigations on the beautiful hamlet of Tintern in Monmouthshire. Years of intense study, radar investigations and archaeological work had proven yet again the correctness of my ongoing deductions regarding the importance of the site. By 2009 my mysterious correspondent was still there – at my elbow! By this time, I was being approached by documentary film makers from America and Canada as well as the United Kingdom. I remained prudent and cautious of stepping into unknown film-making territory, preferring to remain accountable for any decisions I would continue to make in the publishing world. My third book "Kingdom for a Stage" had been published in 2001.[1]

Occasionally, my curiosity would get the better of me and I would attempt yet again to look for a clue as to the identity of my elusive supporter. One day in 2009 I found myself browsing through the presentation copy of 'Festival of Firsts' which was given to every member of the audience at the prestigious opening of 'The Globe' on 12 June 1997. There was a name listed amongst hundreds of others giving credit to those who had played a role in the creation of that 'masterpiece' over the years. The name that seized my attention was that of a board member of the organisation, 'Lord Birkett of Ulverston.' The word 'Ulverston' that day became an irritant. I brooded on it. It was a place name that had figured in the first Barrow correspondence – the geometric drawing representing the 'Hill of Hoad,' Ulverston! I had climbed that hill during my first visit in 1992.

Curious as to the identity of Lord Birkett, I sought out my copy of Barry Day's "The Wooden O." The book had been published in 1997 to coincide with the opening of the 'Globe' and is devoted to giving an account of how the project came about. I had bought my copy later. It now appeared that Lord Birkett had played a major role in certain aspects of policy-making and was chairman of the Artistic Board which had been largely responsible for the appointment of Mark Rylance as the first Artistic Director, a position he held for ten years. I had never been introduced to Lord Birkett or met him to my knowledge. It was not difficult to learn something of the professional life of Michael Birkett. His profile was readily available on the Internet.

My 'executive messenger' had, by his correspondence, become familiar to me through the confidences he shared over those five years. Yet he himself remained an enigma. As silently as my work had come to be treated (and regarded) by the Globe enterprise, I learned as much as I could from records in the public domain and acquired a copy of "The Life of Lord Birkett of Ulverston." Norman Birkett was the father and Michael was the only son of a remarkable Judge, whose eminence is well known for the role he played in the Nuremberg Trials in 1945, as well as other famous cases and legal responsibilities. Norman Birkett died in 1962 and son Michael took over the title.

But, curious as to this tenuous link, I wondered who 'Lord Michael Birkett of Ulverston' really was. He had had a role of influence in the Globe project and that had become something of a question for me. Although he would have known about my research and publications, his position so close to the centre of power would have prevented him from appearing to be an open supporter. Born into a life of privilege and tragedy, destiny had dealt him enormous power in the stakes of positives and negatives. As the only son of Lord Norman Birkett, he had attended the Nuremburg Trials as a youth of sixteen. The trials lasted for twelve months. Later, passionate about the role of the performing arts in society and education, he was one of the prime movers for getting the National Lottery established in 1994. The Wanamaker Globe project was one of the first recipients to benefit from a twelve and a half million pounds award to help its completion.

Thereafter, the wheels of progress moved quickly, providing fresh impetus; deadlines had to be met. Any positive noises regarding my Byrom theatre ideas were silenced. But not quite! I became uneasy remembering those times. I told no one of my suspicions regarding Michael Birkett except Allan my husband, who did not and would not take the possibility seriously. It seemed inconceivable to him that Michael Birkett was behind the odd 'messages' sent to me.

Later he recalled that Birkett had been Producer of the play 'Cyrano de Bergerac' when he was a student at Cambridge University and Allan himself had taken part in the production. Birkett's team included Sir Peter Hall and Peter Brook and their paths had not crossed since. It seemed so long ago; it had to be irrelevant, he thought. The correspondence, as I have described it was coded. The writer was in a way seeking to be helpful. I could not claim he was a 'stalker,' as such. He knew that I would, to say the least, be curious as to his chosen symbolisms. After all, I had worked with Byrom's coded shorthand for years.

I had by this time decided that the sender was also playing some sort of cultural game with me. If the sender truly *was* Lord Michael Birkett, what were his real intentions? Every 'piece' was complex, witty, intelligent and often multi-lingual, a test in themselves, with a boldness which to some might seem a cowardly way of commenting on my dilemma. Twelve years later I sent him a copy of my second book "Kingdom for a Stage" (2001) with a letter apologising for not sending him a copy of my earlier book. I cited a family bereavement (the death of my mother) as the reason for my being less than efficient in sending out complementary copies to those who had played such an important role in founding the new Globe. I knew that his own wife 'Gloria' had died around the same time as my mother. I did not mention the Barrow mail at all. Would he reply?

With a gentleman's courtesy, he replied within days. It was a handwritten letter [2], so I was able to compare the handwriting with that of the Barrow correspondant. Distinctive in style, I was convinced by the similarity. The forming of several letters gave the game away. Also 'Gloria' was a name featured in the first set of papers sent to me. It was quite obscure at the time, yet prominent because of its repetition. I recognised that he knew exactly what he was doing by responding as he did. He included his phone number in his reply, as I had done mine. Although I acknowledged his letter in writing, there was no further contact between us. It had to be that way. In a strange way, we had identified each other now and there was a silent commitment.

The theatre design as implemented was not the original Globe of 1599 which had been hoped for by those responsible for the project. In his letter, Birkett was admitting that fact twelve years later. Successful as a commercial enterprise, and now independent of any funding from other sources, did it matter if it was authentic? It's a question I have asked myself many times over the years. By 2009 Mark Rylance was no longer Artistic Director. He had a successful career in the film industry in addition to his other theatre projects. By this time Birkett himself was no longer a member of the Board. The 'Globe' in London was now established as a centre for Shakespearean heritage and that status would be ongoing. To me, the integrity of Wanamaker's entire organisation had become questionable over time. I developed Birkett's five years of correspondence into a special archive, along with the accumulated study papers which I put together concerning it.

Quite separately, I continued my investigations into the origins of the Brass Plates at Tintern. A plaque attached to the Tintern Abbey wall reads "In 1568 brass was first made by alloying copper with zinc."[3] That was the year the Company of the Mineral and Battery Works was granted a Royal Charter by Queen Elizabeth I. When I saw the list of shareholders in the company, their connections with the Tintern area and Monmouthshire became a significant area of study for me. The list included Sir Nicholas Bacon, the Duke of Norfolk, the Earls of Pembroke and Leicester, Lord Cobham, Sir William Cecil, Sir Walter Mildmay and Sir Henry Sidney. I recognised amongst the list several patrons of the Elizabethan theatre companies in operation at the time. As I investigated further, the interrelationships of these people and place and time and events seemed more than coincidental.

Tintern in Wales is positioned exactly on the border with England, the river Wye being the boundary between the two countries. Estate ownership and certain properties on the Welsh side throughout the centuries became a focus for in-depth study. St. Michael's Church in Tintern and the immediately area surrounding it were subjected to radar scans and an archaeological investigation, commissioned by myself in conjunction with the church authorities.

By the time I received Birkett's letter in 2009 I was well on the way to preparing my next book "The Hidden Chapter" which publicised the investigation. It was released in 2011. As one of the consequences of my research, I had begun seriously questioning whether William Shakespeare was the real author of the plays and works performed and published under his name. My line of enquiry was quite different from those of many Shakespeare scholars and professionals who were asking the same question. In the meantime, I have read many of their works. In my case, the Byrom drawings had been my starting point and had led me this far, so it was to continue.

It was inevitable that I would reflect carefully along the way on the contents of Birkett's letter and previous correspondence, which seemed so palpably relevant to all my inquiries. The implications were of the highest order. He explained in his letter that he had taken on the Chairmanship of the Artistic Board because Sam Wanamaker "asked me to." That was sobering. The Artistic Board during the years of building the Globe were comprised of some of the most prominent names of the theatre world of the day. I knew who they were because Mark Rylance had sent a memo to each member after his visit to me to discuss my research in 1995. It would have been seen by Birkett and distributed with his blessing. In fact, he headed the list of names the memo was sent to. By the time of the memo, I had already received a batch of correspondence from Birkett via the 'postman' in 1992 and 1993.

Rylance's appointment in the summer of 1995 assured him of more than a ring-side seat in the final stages of decision-making concerning the final Globe design, which included the perspective of the Artistic Board comprised of the actors and directors of the theatre world. Such was the case when Mark made sure that I was invited to that special meeting on 23 September 1996. The Artistic Directorate were not happy with the proposed stage design and Tiring House. I described the meeting in chapter three. Michael Birkett would have been privy to and part of those deliberations. He may have been at the meeting; I did not know who he was at the time. Theo Crosby, the architect who died in 1994, had written to tell me that he had incorporated my interpretation of that part of the theatre design into the new 'Globe.' Whether he did or not is a matter for speculation; I am unaware of the ultimate implementation. Being an outsider and a nuisance, I have not been invited to make an inspection.

At the time, neither Mark nor myself knew that Birkett was harbouring and creating a chronicle of sonnet-like verses which were dispatched to me once the theatre was open for business in 1997. His name never cropped up in any conversation over those years or later. Neither did I discuss the 2009 letter with Mark. Birkett's letter was full of the confidence, poise and exclusive knowledge gained from management of business and wide exposure.

By 2009 I was well on my way to completing the next stage on my odyssey – my book entitled "The Hidden Chapter" as mentioned. However, there were challenges to be overcome in bringing it to a close having to do with the administration of St. Michael's Church in Tintern, the subject of the book. Those challenges seemed to have become insurmountable. I was seeking to carry out a second archaeological investigation in a specific area of the church to complete what I considered to be an important cultural enquiry. I suspected that historical personalities associated with the Shakespearean legacy were central to this church and its graveyard. At the time, I was receiving little or no encouragement from the decision-making officers of the Diocese, although for my first investigation there had been an enthusiastic response at every stage. The final chapters of my book were then at a standstill.

Because of what I considered to be tensions surrounding these enquiries within the Church administration based in Newport, and having received the letter from Birkett, I decided to look again at the correspondence from those earlier years. Were there certainties within this 'resource' which could be helpful to me at this time? A man in Birkett's position would not have wasted his time, energy and reputation on idle chit-chat, whatever his method of communication. I noted a 'stature' behind it which caused me a degree of frustration and fear at times. There were 'clues' provided by my analysis of some of the 'pieces,' which at the time I found unnerving, particularly because I had not recognised the 'sender' yet. I had not talked about any of this openly to anyone other than my husband. It might have seemed arrogant and there are always those ready to be critical of an independent researcher. How could *I* have possibly discovered anything of significance? That would have been the question. It is the question.

It had been some time since I last looked at this 'resource' as a collected and stored unit, so I started at the beginning with the letter addressed to "Dear Lady," dated November 1992. There were miscellaneous other papers in the envelope as I said. With the certainty that Birkett was the originator, I approached the exercise differently. The coloured landscape accompanied by the prose captivated me in a different way. The word 'Gloria' was used at the beginning of each rhyming couplet of lines. I realised now that the word was being used as a name. Birkett was referring to his wife Gloria. I noticed too that the same word or name was included in the 'Hill of Hoad' geometric drawing which came in the same envelope. Until I recognised Birkett as the author, I could not have known who 'Gloria' was. Her prominence suggested that she was an integral figure in this unusual and disturbing package of papers. But in what way, I had no idea at the time. I noted the Greek symbols before the signature of this 'messenger' which I had interpreted as 'A Loss For Words.' Meanwhile, I had always regarded the page as an introductory piece and understood 'Gloria' to mean possibly 'in praise of" my book "The Byrom Collection." The title on the page helped to pinpoint the location where it had been posted.

Now curious as to why 'Gloria' had been featured in this way, I wondered whether she was intended to be part of my understanding of his motive for getting in touch with me. I knew nothing about Gloria and there was no reference to her in Michael Birkett's father's biography. It seemed as if Birkett were treating her with reverence. I must do the same. The public record-service came to my rescue. Birkett had married 'Gloria Taylor' on 4 December 1978 in London. She had previously been known as 'Gloria El Fadil,' having been joined in wedlock in London as well to 'El Tahir Elfadil Mahmoud' on 7 December 1963. In 1963 Gloria was described as a 'Boutique Manageress.'[4] When she married Birkett fifteen years later, she was listed as a 'Public Relations Consultant.'[5] These facts were not particularly helpful, so at the time I did not choose to make further enquiries. Gloria died, as noted, in 2001. She was not described as an architect or an archaeologist, which might have provided more clues. The now-known facts did not provide me with a reason to think she could be helpful in solving the riddle.

But, chance or luck, or however you choose to describe it does sometimes drop 'a feather in the lap.' My commissioning Editor, Tom Maschler, had written his own autobiography in 2005[6]. His decision to publish my first two books will always be a mystery to me, especially since he never contacted me after the publication of either book or referenced my work anywhere to my knowledge. His appointed editor, Tony Colwell did, as did Tony's widow after he died. Sometimes I would browse through Maschler's autobiography, reminding myself of this very charismatic figure, founder of The Booker Prize, who was so enthusiastic about my discoveries initially.

On a day when I was perusing his autobiography, I noted Maschler's admiration for the work of the travel writer Bruce Chatwin. So taken was Tom with this talented writer that he let him stay at his country home 'Carney Cottage' for five months while he was writing "On the Black Hill," one of his most successful books. I happen to have copies of some of Chatwin's books in my own personal library, so I took from the shelf Chatwin's biography, written by Nicholas Shakespeare. 'Carney Cottage' is situated close to Llanthony Priory and Abergavenny in Wales, areas which were receiving close attention in my own current research. This Shakespeare's book was published in 1999, two years after Birkett's package of correspondence arrived and the official opening of the Globe. The year 1997 was also the publication year of the paper-back edition of "The Byrom Collection" by Cape, part of the same publishing group as Chatwin's book – Random House.

Glancing through the index of Shakespeare's book, I had what can only be described as an electric shock. There were several references to 'Gloria Taylor.' It turns out that she had been a serious girlfriend for a time and remained a close friend of Bruce Chatwin until Chatwin's own death in 1989 and is given an acknowledgement under her married name as 'Gloria Birkett.' Bruce Chatwin is also the godfather to Gloria's son by 'El Tahir Elfadil Mahmoud.'

Gloria's first husband Tahir was a member of the distinguished 'Mahdi' family. His grandfather was Siddig El Mahdi, a former Prime Minister of Sudan in the 1960's and then later in the 1980's. Sudan acquired independence from Britain in 1956. Bruce Chatwin visited his former girlfriend, now married in 1965 to Mahmoud and living in Khartoum. She lived with Tahir in a flat with a balustrade terrace next to the Mahdi Palace. Bruce arrived on 5 February and at that time Gloria was pregnant.

He stayed with them for one week sleeping on the terrace. Not too long afterwards, political difficulties persuaded Gloria to return to England. This was after the birth of her son.

The shock to me on reading this entry was the mention of the 'Mahdi' dynasty. Why? Because the use of the word 'Mahdi' had been included in one of the pieces of correspondence from Birkett.

On receipt of the 1997 package, as already mentioned, the author was still a mystery to me, so I could not understand the connection. Now the veil was gradually being moved to the side. The piece in question was one of the twenty-one pages and bore the title 'The Border Reivers.' It was concerned with the seeking and exchanging of rarefied knowledge. The implication was that two of my 'consultants' were aware of some of the rarefied knowledge contained in "The Byrom Collection" and were looking for contacts who might help in areas that they were not able to cope with. This had come to Birkett's attention via his own 'Mahdi' connections and caused him some irritation. On the reverse of the page 'P.T.O.' were written the words 'Mahdi, Mad he, Made he.'

The two consultants who were helping me at the time were Leon Crickmore and the Rev. Neville Barker Cryer. Birkett apparently thought them naïve. I had found this piece intimidating for some years because it seemed to have been written from a position of 'knowing,' although I knew not how. My fears were not lessened when I recognised that there was a close connection between Birkett and Gloria's families and Tom Maschler, my commissioning editor.

Another page of the twenty-one page grouping had already caused alarm for years because the implication was that even before the publication of my first book in 1992 there had been 'dealings' at the planning stages of the Globe project – 'wheels within wheels' – with money changing hands, associating Tom Maschler or his editorial staff with some of it. There was also the suggestion that I and my theatre work would be treated with silence by the media and others.

For a long time, I had thought maybe I was mistaken in my interpretation of the drawings and doubted myself. Again, I said little to anyone. On reflection, I think that it was I who was naïve – in doubting my own understanding and not suspecting the machinations which seemed to surround my efforts to reveal what I had found. As I looked at this page even more closely, nuances in detail became stark. There was serious intent behind the writer, this 'messenger.'

The fact of Gloria's contacts in Cairo and elsewhere may have been alarming to some who were in a position to suppress my work with this enigmatic collection of drawings. The learned scholars known to Gloria may have been able to interpret the methods used in their geometric

The Lieder Laud and the Jolly Mullah

construction and might have been able to understand my workings of them. That possibility may have been a threat to some who possessed power and wanted to protect it. Who would that be? Birkett wished to convey that understanding to me and did. I quote from one piece in particular:

> From neath these bunds of sad decay
> I fancy that some voices say
> "once flesh and bone and blistered fingers
> we bide awhile, our spirit lingers
> wishing to see what we have started
> before we are, truly departed
> and left you here a base to build
> lift up your eyes and be fulfilled
> the bright light beakons as it ever did
> build up build up, the pyre amid
> distributed seed and sentient spark
> follow the way of The Knowers Arch
> Helix shows the chosen way
> God's gift is yours we've had our day
> Get off your knees and get thee hence
> reach for the stars…it's only sense"
>
> Solitary I saw the sunlit road
> my chalice filled…and overflowed

These lines are numbered 54 to 71 of a piece 102 lines in length, a highly significant number. The piece was entitled 'Hymn of Furness Abbey,' handed to me personally by the 'postman' in Barrow on that February day in 1993. At this point there were two bays of the 20-sided new Globe theatre standing in London. By that time, I realised that my interpretation of an eight-sided building had been rejected. I have quoted lines 91-102 of the same 'piece' in chapter one. With those lines the author was suggesting that we meet, yet there was too much mystification for me. It did not happen.

Years later, I see the lines as allegorical in their use of language and phrasing. I leave the reader to interpret them as seems fit. However, I feel obligated to draw attention to the use of two words. First, 'Helix' stands for an object having a three-dimensional shape. I understood 'Helix' in Birkett's piece to represent the key drawing in Byrom's 'Globe' sequence, the three-dimensional architectural 'plan,' including horizontal and vertical measurements, clearly composed and cleverly executed. 'Bunds' I interpreted as boundaries. I also understand that the last two lines quoted mean that the author understood the 'truth' as an individual and was overcome by the knowledge.

One hundred and two is one of the key dimensions of the original Globe theatre (1599), as I have come to understand the design. It is a very complex matrix. Fortunately, I had also recognised that the accompanying drawings in Byrom's collection should be used with the 'Globe' drawing. In the Globe drawing, there are actual cuts in the card, carefully indicating that it has three-dimensional instructions embedded within its geometry. The brass plates from which these details were taken I later recognised in the Science Museum's Boyle Collection of Brass Plates in London.

Over the years I have consulted many experts in the field of mathematics, including members of 'The History of Mathematics Society,' to gain a better understanding. Where relevant, these experts have been quoted and examples of their work included in my earlier books. Birkett's correspondence was different from the contributions others made. It was exclusive and in its own way demanding. The (Globe) piece 'A Truth Lies Standing,' already discussed in Chapter 6 contains the lines:

> So look behind the hard façade disclose eternal treasure
> talk is cheap and so is verse unless you know The Measure
> truth and falsehood intertwined and in here I do ensconce
> the art is how to weigh and weight each word and speak both
> at once.

He goes on to invite me to meet him again. At this point I still did not know who Lord Birkett was and his connection to the mysterious correspondence. He goes on to say that if I chose not to, then I should visit 'The Prince of Wales School of Architecture' in London ("Wheels within Wheels"). This establishment had already been presented with a copy of "The Byrom Collection" by me at their request, via my publisher. A representative of that group, Keith Critchlow, had been invited to the launch of that book at Groucho's London in 1992. If the school had wished to discuss my work further I would have made myself available. There has been no such invitation, meeting or contact.

The effort of dealing with the interpretation these enigmatic 'pieces' sent to me was very demanding and unforgiving. I did not understand the reasoning or logic behind such mystification and felt that I could not be a part of the charade actively. It did not matter how well-meaning the motive might have been. As far as I was concerned there was too much anonymity regarding the whole process, despite a signature or even the lack of it. Birkett's letter of 2009 altered that perspective in part. But so many years had come and gone. Now reflecting on the 1997 Globe publicity programme, he was advertised as 'Lord Birkett of Ulverston.' In the 1992/1993 correspondence, with the mysterious drawing of 'The

Hill of Hoad' Lighthouse in Ulverston and the name of 'Gloria' (Taylor) carefully positioned where it was, seemingly central to the geometry, it seemed Gloria's past personal connections might have played a role in some way.

And yet Birkett in one section of the long-quoted piece describes himself as 'Herr Schneider' a 'Tailor's Dummy'. For the record, Birkett's grandparents owned a well-established draper's shop in Ulverston and his father Norman had served a seven-years apprenticeship as a tailor with the firm before going to Oxford University. Another clever turn of phrase with a questionable meaning. Is this how Michael Birkett saw himself? Still, I recognised that the preparation of some of these pieces must have taken a considerable amount of time and effort to create. Gradually a silent respect was developing in a way that I had not anticipated.

Who were these individuals? How did they spend their days? Birkett clearly thought that I would understand. Apparently, they were concealed supporters and influencers prepared to support my work but I did not take the vital step. After 2009, as time went on, I regretted not finding a way to take up the cues that were being put before me. Now recognising a sense of purpose from Birkett towards my own research, I looked for credible justification for his interest. I quote an example from "The National Theatre Story" by Daniel Rosenthal published much later in 2013. Covering the years 1848-2013 his studies are extensive and detailed. In theatre 'gala' events taking place at 'The Old Vic' and 'The National Theatre' in London in 1976, Rosenthal commented:

> Michael Birkett's peerage and immaculate Establishment
> Contacts made him the obvious choice to manage gala
> protocol.

Birkett managed the gala at 'Plunder' at The Lyttleton on 15 March 1976 at which Princess Margaret was present. On 25 October 1976 there was another gala performance to which the Queen, Prince Philip and Princess Margaret were to attend along with other dignitaries. Sir Laurence Olivier was starring in 'Jumpers' but was in no mood to join in the celebrations afterwards. The Royal party having left the theatre, Rosenthal records the event as follows:

Olivier:	Michael can we get out of here? Birkett; Out of ....?
Olivier:	(gesturing around The Lyttleton foyer) Out of all this.
Birkett:	Well you can come up to my office. I've got this gorgeous girlfriend with me, I've got a cocktail cabinet, and what's more I've got a fridge with ice and vodka.
Olivier:	Let's go.

The quote continues as follows: "Birkett's girlfriend, 'a devastating lady,' the most talented, beautiful thing you ever met" was Gloria Taylor, a Christian Dior model-turned-publicist for the Royal Court. The cabinet might have been a prop from 'Private Lives' with inlaid wood, the top flipping open to reveal glasses and cocktail shaker, imported by Birkett from his father's golf club. The trio drank for hours. When Olivier rang the next day to thank Birkett for receiving him, he asked after 'that adorable girl' and was told "Larry, she is mine."

Birkett married Taylor in 1978. Rosenthal's 'gala' account of the celebrations after the Royal visit to 'Jumpers' and Laurence Olivier's reaction that evening was very helpful to me. Rosenthal's book was not published until 2013, so I could not have been privy to such information unless I had been told personally by someone involved at this colourful event around that time. Gloria's glamour was memorable and somehow Gloria was central to this mystery.

Birkett's largest batch of correspondence arrived in 1997, twenty years after that event. The play 'Jumpers' is about a team of professors with absurd differing views on philosophical issues. Perhaps that is an over-simplification, but it is enough for my purposes in that the author, in one of the twenty-one pages (Birkett) entitled 'Rev on nous mute ons' includes the following line:

Setting his teeth into masses of woholy thinking jumpers

My interpretation of this line had been, all those years ago: "Giving his ultimatum to the group of woolly (lacking in substance) thinking experts." I deduced from that that Sam Wanamaker had strongly urged the academics of the Globe project to 'get on with it!' The use of the word 'jumpers' tied the author of this mysterious correspondence more closely to Birkett. The 'gala' incident had made a lasting impression on him. Gloria's affect and influence on Birkett's life after their marriage two years later must have been dramatic. Her accumulated knowledge and experience of the culture of the Middle East would be with her for life. Contacts in connection with Temple architecture must have been available to her. I think she was using them as a resource which was then made available to Birkett. Birkett clearly had an interest.

The collected correspondence, as time went by, became more and more special to me. Occasionally a word or phrase previously unclear would stir in my memory an event or an incident that had been brought to my notice in some way. Gradually I came to learn and understand the culture and history of the world of theatre in a way that had not been necessary to me before. My insights into Byrom's collection of drawings had led me in that direction.

Birkett was a foremost figure of the theatre and media world of the time, as well as having a wide knowledge and understanding of 'esoteric' subjects. The latter was not a necessary component of my research at the time although I recognised his interest. That situation has since changed. Although I welcomed his anonymous acknowledgement of my work, it remained a private matter. I later understood that the complexities he had inherited in his youth carried heavy responsibilities and perhaps an imposed loneliness. In some ways he became a solitary man when he inherited his father's Peerage in 1962, when his father died suddenly during a cardiac operation. There was no funeral and no commemorative service. Norman Birkett had left instruction that Michael (who was a young man then) was to be present at the disposal of his body for cremation. The responsibilities that young Birkett inherited at that moment would have become very clear to him. Still, his acquired Peerage also carried extra powers and connections as well as responsibilities.

Circumstances were such that I was now focusing on my own developing thesis on Tintern. I was totally independent. My research was my own responsibility. That too was a lonely position to be in. Maybe he recognised that. In his later lengthy correspondence to me, after I had identified him as my 'executive messenger,' Birkett talked of his own aspirations. For a time he thought he would study for the Church, but then admitted that it would make him more cynical. He also thought to study and teach 'Middle Eastern Deities' which would help in his understanding of the principles of sacred geometry as they applied to the Collection. Eventually, he wished to know who had ultimately decided to ignore the merits of the 'Hermetic rule of number' as used in the Byrom drawings, which should have been applied to the new Globe. Maybe he did know but chose not to record the names of those responsible. He was obviously very supportive of the initiatives I had taken and was attempting to be encouraging. There were other facets of his life and personality that he chose to describe in his correspondence. They are not relevant to this narrative so they are not included here.

It was now imperative that the investigations and findings which were centred round the church of St. Michael's and its graveyard and surroundings should be published. The radar and archaeological data and artefacts discovered during the work had to be placed on record in the public domain. That was part of my agreement with the Church of Wales administration.

My book "The Hidden Chapter" was launched in Tintern on the second and third of April 2011. Celebratory events were arranged on both days. On Saturday the second of April invited guests gathered in the church for a résumé of the results of both the radar and archaeological work carried out by the appointed professionals. This was followed by afternoon tea at the nearby Wye Valley Hotel. At that venue the contributors to the book were present as guests for a pre-arranged dinner on the same evening.

The following day, Sunday the third of April, the owners of 'The Nurtons,' the nearby estate which figures so strongly in the history of Tintern, hosted a magnificent event in celebration of the publishing of the book. Guests from all over the country enjoyed the home-made delicacies, music by the well-known Welsh harpist Claire Hamilton and talks relevant to the publication. I had included all those who had played a part in the creation of the book which would now be taking on a life of its own. But there was one person who could not be invited to the celebrations, Lord Michael Birkett. It would have been awkward. Discretely, I acknowledged his attentions with a quote from his 1997 correspondence. In my book the section is called 'Some Reflections.' His 1997 quote:

> Some things come and some things stay, and others must always
> pass. These words were spoken by an Astrolabe, sung by a Tongue
> of Brass. Those mighty seventy scholars high in the tower of Babel,
> they had things firm, they knew each term, of a chorus Astrolable.

He was, in my opinion, referring to the Globe academic board. I included an acknowledgement to the Author: "My earnest correspondent remained hidden behind the veil of pseudonyms and I have spent many sleepless nights blessing the silent rhetoric of his anonymity, but I salute his courage." I sent Lord Birkett a copy of my book "The Hidden Chapter" and never heard from him again.

Lord Birkett died on the third of April 2015, the anniversary of the launch of the book in 2011. It was now certain that I would not be able to meet him about his disclosures to me and for a time I was saddened by this empty space in the enigma. Although I was by then certain he was the originator of this 'resource' and had read his 2009 letter which confirmed his identity and intent, I realised there could be denials by 'doubters' if I were to release that information at large. I felt that his own concerns for Globe heritage had been thwarted, as had mine and those of giant literary intellects, for example, Sir Peter Hall, also mentioned in his letter. Birkett's hopes for my 'ideals' had been disappointed.

I wondered who else knew of his opinion regarding my conclusions about the original Globe design, but felt that given my own silence for so many years about this cache of correspondence, it would not have been appropriate to begin asking questions in actual 'Globe' circles. I had decided to wait until documentation in a recorded form was available in the public domain. Maybe there would be a confirming clue somewhere. I was lucky. Lord Birkett referred to connections with the 'Mahdi' in official documents and that is enough for my purposes. He was the source of the

'Barrow' correspondence, erasing any remaining doubt. Left still in doubt is the 'Source' behind the 'source,' a matter still under investigation.

And what of his willing 'postman,' Brian Charles, who committed suicide at the end of 1997? He described himself as a loner during our meeting. He clearly had to be known to Birkett in some intimate way; maybe he was a hidden sibling or other family member. His father Frank Charles was well-known in Barrow but died young in 1939 in a flying accident. Brian was five years old at the time and was brought up mainly by his grandmother Edith Searle. His mother remarried and moved away. A photograph of his celebrity dare devil father shows him to have a shock of red hair. I have recorded in my account of meeting Brian Charles in February 1993 – how he sported a toupee-style of red/brown curls.

Lord Norman Birkett was called 'Carrots' and 'Copperknob' by his teenage friends at the local grammar school in Barrow [6]. He had a mass of sandy-red curls which he had inherited from his father. He died on the tenth of February 1962. Lady Gloria Birkett, whom I had come to consider one of my hidden champions, died on the same date as her father-in-law, thirty-nine years later in 2001. My Barrow 'resource' remains intact and complete. My research programme continues. The Tintern findings have encouraged me to look even further into the property holdings and families who have occupied them for centuries. The Shakespeare authorship question took on a new dimension that I could never have expected. For what I have come to believe in and respect – for all of those concealed champions – I felt that I had to continue. It was an awesome task.

CHAPTER EIGHT

Tintern Anomalies

Since recognising the 'measure' and numbering system in the geometry of the Byrom drawings and brass plates, my life's sense of purpose changed in some ways. The truth and simple elegance of what the number nine represents, and then seventy-two, and then one o' two, gave me a different sort of clarity. My interpretation of the original 'Globe' design was based on those perceptions. That would not change. The truth in the design was evidence of perfection through numbers. Once recognised as a numerical discipline, it affected me in several ways, even in my day to day routines and general understandings, boosting a confidence which I could share with but a few. I did not plan it that way. Recognition is singular to the individual and is unique in each of us according to our reaction to what we observe at any given moment. I have learned to respect that quality. When I first came to handle the drawings and to explain them to others, reactions were quite different, and sometimes created sensitivities. It was as I saw it nothing to do with natural intelligence.

I came to regard myself as privileged in having such an opportunity. John Byrom had preserved a unique approach to numbers and the special knowledge inherent in that approach as presented in the Collection in his own day, which has been in turn handed down to the inheritors of his estate. For a long time I was not at liberty to know just how rarefied the assembled package was. My books have been stepping-stones along the way of my encounters – the findings and conclusions thus far. As noted, my book "The Hidden Chapter" was written as a result of my research into Tintern in Monmouthshire in Wales and the surrounding area. Seeking the provenance of the brass plates had led me to Tintern and the brass works, which led to investigating the luminaries of the sixteenth century who had established the brass works there, including the Francis Bacon family, which in turn led to me investigating Francis Bacon himself. Some scholars have claimed that he in fact is the real author of Shakespeare's plays. I began with serious doubts about that claim yet kept an open mind. I would do my own investigation. It seemed all was coming round full-circle. I needed to know, to learn as much as I could about the drawings and the brass plates.

St Michael's Church, Tintern

At the time of that 2001 publication, there were other mysteries which had by no means been answered. If anything, any answers that I had been able to suggest in my book only deepened the enigma. The Church of St. Michael's in Tintern seemed central to the mystery, so prior to the publication of "The Hidden Chapter" I submitted a formal application to carry out further archaeological work in the churchyard. Permission for a first investigation of that kind had been granted prior to writing the book. I would be seeking to establish whether there was the possibility that Francis Bacon had been laid to rest in this little church, and even kinsmen of his. Francis Bacon had been a shareholder in the brass works and had many connections to the area. I had reason to think that his burial there was a likely scenario. In addition, I suspected that there could be artefacts buried underground which might contribute to answering the Shakespeare authorship question. I wished to confirm whether there were underground vaults in the church itself, other than burials in the churchyard, which had previously been implied by my commissioned radar investigations. Those findings had already identified for the first time that Roman 'workings' had been carried out nearby and even on the site.

I waited for three years for an official response from the Church administration and was seriously concerned by the silence. I had received no decision one way or the other to this Faculty Application. I discussed the matter with one of the Church Wardens, Andrew Reid, a local resident, who after careful negotiations was able to arrange an unofficial meeting with Chancellor Price, the legal church representative of the Diocese. The meeting took place in the church itself on Wednesday 25 June 2014 starting at ten o'clock. It was one of the most difficult days I had ever had regarding the project. My diary entry for the day records the session as follows:[1]

Present: Chancellor, Tim Russen, Registrar, Peter Webster DAC, Archaeologist, Andrew Reid, Alan Carter, Wardens, Nora Hill, Rector. Member of the community. Chancellor chaired the meeting, Russen acted as recording secretary. Chancellor began: Outlined situation regarding Faculty Application – covered altogether period of 15 yrs. Faculty App. normally looked at two facets, spiritual side – the considerations concerning whether there was a current spiritual association/contribution, and/or the history side of application. Weight: towards spiritual, but he tended to look at both equally. He said this application has far-reaching consequences, very wide-reaching, very complex; included wire works/brass works. He covered the contents generally very well; clearly, he had read the book. I thanked him for that and then asked how confidential was this meeting; where would this info. finish up, purpose etc. He said he wanted to set a date for the 'Consistory Court' session and that today was to determine that, or closure.

It was up to me how much I disclosed, no confidentiality, would need to consult. I would have to consider all of that.

One of the difficulties I found presented at the meeting was that some significant findings in the Russen archaeological and radar reports had not been seen by Peter Webster, the D.A.C. archaeologist present or Tim Russen, the solicitor Registrar. The Chancellor stated that those needed to be studied along with any new research findings I had made since the publication of "The Hidden Chapter." A list of items missing from my application, which could not be located in the Church Administrative offices, would be sent to me directly, so they could be submitted again by me for consideration by the necessary personnel, and then a date set for a 'Consistory Court' session.

By the end of the meeting I understood why it had been arranged to take place in this way. My application had "far-reaching consequences." The necessary appointed professionals had not seen, before this meeting, the 'findings' which would have supported my real interest in investigating further. It was also necessary, it seemed, for me to submit further evidence to that already published in my latest book three years earlier. As noted, my Faculty Application had been submitted before that publication.

The Chancellor had done his homework, read "The Hidden Chapter," and after the meeting studied and discussed with me the facsimile of the first edition of Shakespeare's folio of plays (1623). I had brought it with me. My personal opinions on the edition's original presentation and Chancellor Price's interest gave me an opportunity to say so later. His agenda had not allowed for the topic to be included in the meeting itself.

On Saturday, 12 July a letter was posted to the Diocesan Registrar's Office in Newport. In it, I withdrew my Faculty Application altogether from further consideration. I had deliberated over my final decision for three weeks by discussing topics covered in the meeting with consultants who had been involved in the publication of my book. My diary entry summarises the overhanging sadness of the day of posting the letter:[2]

> So! Today it was the turn of the Chancellor and the Registrar. My letter of 'withdrawal' was to be directed to them. In principle I used the same letter as that written to Andrew Reid (and posted yesterday). It meant that it was partly a case of 'cut and pasting' from Allan's computer. Of course, there were additions and specifics. I found it all very difficult if not upsetting and I was glad when it was all finished. I corrected the draft copies a couple of times, but in the end the document looked very well. Allan was able to get it in the post for 12:00 noon, and so it should be in Tim Russen's Office on Monday morning. Thinking it all over – the implications – it has been a shattering experience and I am not sure how the Chancellor will respond. It

has become clear that since Janet Bone's letter of 18th March 2010 (D.A.C.'s decline of approval 'you will be hearing from the Chancellor') there has been nothing really until 25th June 2014. Four years! It seems that the new book had not affected any decision from the Chancellor one way or the other.

My decision had been influenced by the collective response of my consultants who spoke with one voice. My diary entry for 29th June records their reactions:

> Generally speaking the format was the same and the story did not improve with the telling. Whatever I said the reaction was the same. The 'Church' cannot cope with the implications. It is overwhelming and too much. Michael Darlington (a former Registrar himself for the diocese of Manchester and my solicitor) could understand the format, and of course the Chancellor was under-prepared and would not be able to move forward with a date. The negligence of the administration is almost too much to bear. Gradually the realisation is settling in my head. Neville always said that there was nobody there capable of understanding.

I valued particularly the opinion of the late Reverend Neville Barker Cryer, who whilst I was carrying out most of my research, was the Secretary of the Quatuor Coronati Lodge, the leading research Lodge for the Freemasons in the United Kingdom and beyond. He was also the Provincial Prior of the Knights Templar of Yorkshire. He had recognised aspects associated with St. Michael's church and the graveyard which had specific associations with symbolism connected with Freemasonry and the Templars. He and I had regular meetings to discuss my research.

In withdrawing the Faculty Application, I felt that I was relieving the church's 'decision-making' process of needing to discuss or consider my research further. The church itself was going through changes of personnel. My application only added to the problem. This contributed to my decision, but I considered that my enquiries were serious and justified, so the church issues would be put on the 'Pending' list.

Fortunately, the established recorded data in the radar reports and archaeological findings, when put alongside my research work carried out since the publication of "The Hidden Chapter" in 2011, made my disappointment temporary. I turned all of it into a workable strategy of my own. I could not allow the church's dalliance in dealing with implications of a historic nature by proper investigative procedures get in the way of progress.

The nearby estate of 'The Nurtons' was very much a part of my investigation because for centuries there had been a close connection between the estate and the church. It was only over the last hundred years that there had been the need for a separate and different approach when looking at this particular area for research purposes. In part, a major through-road and the introduction of the local railway had forced developments which changed the cultural environment of Tintern. The River Wye as a means of transport and local fishing became more limited.

The natural charm and careful nurture of this beautiful spot was now affected by modernization. The traditions which had remained insular through the centuries were being forced in certain ways to move with the times. This inevitably caused a change in the ownership of sites that had remained with the same families for generations and an influx of fresh blood with different ambitions forced the need to develop amenities. Tintern Abbey became a site of Welsh National Heritage and much of the surrounding area was sold by the Duke of Beaufort to 'The Crown' at the beginning of the twentieth century.

Fortunately, the owners of 'The Nurtons' as current custodians of the historical nature of this significant part of Tintern's past have preserved the integrity surrounding its history. They have carried out research themselves and have generously made their findings available to me. I in turn have deposited with them copies of the radar and archaeology reports carried out at St. Michael's. This arrangement has given me the opportunity to study source material, much of it original, using as tools the Byrom collection of geometric drawings. I have used this resource to collate the data from my earlier investigation carried out regarding the provenance of the brass plates.

I was fortunate to have another 'tool' at my disposal. Having withdrawn from the second Faculty Application, I felt free to relax the restrictions of protocol which were in place while the application was under consideration. The Chancellor had influenced my decision by saying that my research had "far-reaching consequences." But the consequences unknown to him had more than he imagined as an additional strand attached.

The links which were being forged as a result of my investigations were having a profound effect on me personally, for my father was born in Whitelye, Tintern. For generations his ancestors had been very much a part of the community life of the area. For instance, he and his father (my grandfather) had discovered Bronze Age tools in a field while working for The Forestry Commission at Livox Farm, not far from the Abbey, in the 1920's. The historic find is part of the Bronze Age Collection at Cardiff Museum. My grandmother was the organist attached to St. Nicholas's Church, Trellech for forty years nearby. Her parents were responsible for financing the building of the protestant chapel in Whitelye, now a private residence. My grandfather sent his three sons to study at the Duke of

Beaufort's forestry school in the Forest of Dean. Taking it further back, my great-grandfather was a licensed salmon fisherman on the River Wye. I will stop there. My family's descendants have moved away and there are but few family connections in the area now.

I started my education in London where my father held a position with the London Parks Department. For many years until early adulthood, family holiday time was spent in this special place. The families were very much part of the community. When I found a need to investigate the brass works at Tintern twenty years ago it was required to go there and any family associations were irrelevant. But during those years between then and now I was able to draw on rarefied additional information with an sympathetic eye and ear. It has convinced me that there are profound mysteries, yet unresolved, in connection with this area of natural beauty, with its centuries of history and elegance.

The solemnity of not knowing can fire the imagination to look a little further and that is what happened.

CHAPTER NINE

Surprises and Family Links

A<small>ND SO IT</small> was that during one of my lengthy periods of research into the Tintern area I was taken by a cousin of mine who lived there to look at 'Livox Farm' and the spot where my father and grandfather had discovered Bronze Age axes while digging the foundations for a boundary fence. My grandfather worked for the Crown Forestry Commission at the time. My father had not yet started his course of study at the Duke of Beaufort's forestry school but was working with his father on that day.

The area of Livox Farm was among the lands given to Tintern Abbey in AD 1131 by Walter de Clare and had been farmed for generations. The discovery of Bronze Age tools there was regarded in the 1920's as unusual. It is not every day that such items, dating from between 1200-3300 BC, are unexpectedly discovered while carrying out a seemingly straightforward job in a field. It became a favourite anecdote of my father's on family occasions when matters of local interest came up.

So on the day of my visit to the site I was taken on a tour and by arrangement visited the local church, another by the name of St. Michael's – of Michel Troy. The church was thought to have origins dating back to 1208 and was built by the 'Clare' family, lords of Usk and Trellech. Walter de Clare, as stated, had been the founder of Tintern Abbey in 1131. His father Richard had come to England with William the Conqueror.

In writing my diary account for the day, as well as noting points of interest in connection with Livox Farm, I recorded my search for background historical facts and individuals associated with the immediate area and the church I had visited. Were there any dignitaries I would recognise as having connections to the study in which I had been engaged for many years? To my surprise the Rector of the church from 1602 until 1611 was Nathaniel Baxter, at one time classics tutor to Sir Philip Sydney. His sister Mary was the mother of the 3rd and 4th Earls of Pembroke to whom the folio of Shakespeare's plays was dedicated in 1623. This Pembroke family were the 16th and 17th century patrons of the St. Michael's Church in Tintern where I had been carrying out my investigations. The 2nd Earl of Pembroke died in 1601, the year before Nathaniel Baxter took over the responsibility for the nearby church, St. Michaels' of Troy. Close to that church was 'Troy House', not far from Livox Farm. These places are of special interest to my research because of their links to John Byrom, John Dee, Shakespeare and the 'hermetic' knowledge contained within the Byrom collection of drawings which are the subjects of my previous books. John Dee in particular was a dominant figure around whom much of my later research centred.

Troy House was a mansion built by the Herbert (Pembroke) family. The 1st Earl of Pembroke's illegitimate son, Sir William Herbert lived at Troy House with his second wife Blanche Milborne. They married sometime between 1500 and 1502. The husband died in 1524. By the 1530's his widow 'Blanche,' whose official title was Lady Herbert of Troy, is recorded as being of the "Royal Household of King Henry VIII" and became part of the nursery staff of his son Prince Edward, later King Edward VI, and daughter Elizabeth, later Queen Elizabeth 1st.

Lady Herbert of Troy's niece, also named Blanche (Parry) joined the Royal Household at the same time as her aunt, at the age of twenty-five. This Blanche (Parry) became part of the nursery staff caring for Princess Elizabeth and her brother Prince Edward. As the years went by, she became Elizabeth's closest confidant. In 1565 Blanche was appointed the Chief Gentlewoman of the Privy Chamber. This responsibility gave her control of who had access to Elizabeth when she became Queen and put her in charge of her jewels and furs. She was also responsible for looking after her private papers, wardrobe and library of books. Blanche was considered part of the household until her death on 12 February 1590. She was buried at St. Margaret's Church in Westminster and the Queen paid for the funeral.

Blanche Parry had been a constant companion to Elizabeth during Elizabeth's childhood and remained so for much of the Queen's turbulent reign. Her loyalty was exceptional and her power base critical in the Queen's day-to-day routine. Recognizing that it was her aunt, Lady Herbert of Troy, and her connections with the Pembroke dynasty that had provided Blanche with the initial Royal entrée, the Tintern personalities of the time became even more intriguing, as did the area itself. Prominent families were bound by a closeness evidenced by the juxtaposition of properties and their ownership, as well as their professional responsibilities and patronage within the local churches and 'shares' in the brass works.

Although the main strands of family histories and their responsibilities had been included in my previous studies, I find it necessary now to note further instances of some significance. Lady Herbert and her husband Sir William Herbert need more attention in the narrative. They had hosted King Henry VII, father of King Henry VIII, and other aristocrats at their Welsh mansion Troy House in August, 1502 shortly after they married.

Henry VIII succeeded his father in 1509. Before marrying his second wife Anne Boleyn in 1533 he created her Marchioness of Pembroke, having already given her as part of her dowry the Lordship of Usk, which included St. Michael's Church, Tintern. Anne Boleyn's reign was short. She was beheaded three years later in 1536, allegedly for marital unfaithfulness. As a child of three, Elizabeth the future Queen of England, was now without the guiding influence of her mother, Anne Boleyn. Those who participated in her daily routine would inevitably become substitutes for her mother, of one sort or another. Lady Herbert of Troy retired from royal service with a pension in 1546 when Elizabeth was in her teens and thereafter died before Elizabeth's coronation in 1558. Blanche Parry continued in service.

But, intrigued by this Herbert connection, I recognised another historical fact which was to become pivotal in my quest for evidence to help solve the still unanswered questions. In 1552/53 John Dee, the noted astrologer, entered the service of William Herbert, the first Earl of Pembroke. This William was the grandson of an earlier Earl, who had been executed in 1469 and was the father of Lady Herbert of Troy's husband. He was buried in Tintern Abbey.

John Dee would have been twenty-five years old when he joined the Herbert household and was already a highly respected scholar with appointments to other households close to the Crown. Those connections were to continue throughout his long and eventful life. I had been studying his diaries for many years. It must be remembered that five hundred years ago travel was limited and the technology providing transfer of knowledge was restricted to books, manuscripts, paintings and letters. Word-of-mouth was the means for keeping up to date with contemporary matters and an exclusive occupation for those who specialized in doing so. Direct contacts between families was therefore of great significance and the written records of such even more so.

The Fenton edition of Dee's diary, as it has been published, begins on 16 January 1577. (Dee's entries prior to 1577 are limited to a few between 1547 and 1554 added in an appendix). The 1577 first entry reads:[1]

The Earl of Leicester, Mr. Philip Sydney, Mr. Dyer etc.

I have always found this entry of interest because of the group of personalities involved. They were on their way to visit John Dee at his 'Mortlake' house, just outside London. The Earl of Leicester was Robert Dudley, whose relationship with Elizabeth beginning in her teenage years has been well-documented by other observers over the centuries. Mr. Philip Sydney's sister Mary married the 2nd Earl of Pembroke. Sydney was also the nephew of Robert Dudley, and his classics tutor, Nathaniel Baxter, became the Rector of St. Michael's Church, Michel Troy, some twenty years later. By that time both Philip Sydney and the Earl of Leicester, Robert Dudley were dead. And what of Dyer? Edward Dyer, an Elizabethan courtier, was born at Glastonbury in 1543. He is described by the author of "The Queens and the Hive" Edith Sitwell as an "agent of Queen Elizabeth." But there is more to learn of this gentleman.

The January entry of 1577 in Dee's diary served to create another purpose for me. Just a few months before, in 1576, the first Elizabethan playhouse was opened in Shoreditch, London, aptly named The Theatre. Years later it was taken down and the timbers were born across the River Thames and rebuilt as The Globe in 1599. As a result of my investigations, it is my opinion that John Dee was responsible and central to the design concept of the building. Edward Dyer died in 1607, John Dee died one year later in 1608.

Doctor Dee's scientific and mathematical achievements have been scrutinised by various experts for many years. Edward Fenton's edition of Dee's diaries, which were published by Day Books in 1998, are very useful as a resource. He has competently decoded the complexities of Dee's symbols, added to Dee's entries for record-keeping purposes.

So, at this point we must look at the diary entry for 16 July 1579, the significance of which deserves to be noted:

> Arthur Dee christened: Arthur Dee was christened at three of the clock afternoon. Mr. Dyer and Mr. Doctor Lewis, Judge of the Admiralty, were his godfathers; and Mistress Blanche Parry of the Privy Chamber his godmother. But Mr. John Herbert of East Sheen was deputy for Dr.Lewis and Mistress Awbrey was deputy for my cousin Mistress Blanche Parry.

John Dee, in this 'christening' entry in his diary describes Blanche Parry as his cousin. Well over the age of sixty-five and still in the employment of Queen Elizabeth, she was to become the godmother of Dee's eldest son Arthur. Her appointment would be regarded as one of privilege and an honour. Arthur was John Dee's first born and Blanche was by this time one of the Queen's closest confidants, having been in her employment for over forty years.

The realisation that Blanche's family had connections with the Tintern area had heightened my decision to study the entries in Dee's diary from around this time. Queen Elizabeth would have been aged forty-six years on the occasion of Arthur's christening. I had noted that there had been some to-ing and fro-ing of Dee by invitation to visit the Queen at Windsor Castle a few months earlier and then later at Richmond Palace. At this time in 1577 and 1578 discussions were under way concerning the Queen's 'title' in connection with "Greenland etc., Estotiland and Friseland." Dee had been engaged in plotting navigational routes to foreign lands. The entries which followed from September 1578 became of special interest. I include the block of entries below. It will be seen that the record has gaps of days and weeks, which suggests that it was the incidental timing of events that occurred which needed to be recorded when Dee entered them:

1578	
24 Sept.	The first rain that came many a day: all pasture about us was withered.
25 Sept.	Rain after noon like April showers. Her Majesty came to Richmond from Greenwich.
8 Oct.	The Queen's Majesty had conference with me at Richmond in-ter 9 et 11 antemeridian.
13, 14 Oct.	(Elizabeth) hor. 11 ante meridiem
16 Oct.	Dr Bayly conferred of the Q. her disease.
22 Oct.	Jane Fromonds went the Court at Richmond.
25 Oct.	(Elizabeth) a fit from 9 after noon to 1 after midnight.
28 Oct.	Hor. 5. The Earl of Leicester and Sir Francis Walsingham, secretary, determined my going over for the Q. Ma(tie).
29 Oct.	(Elizabeth) hor. 5 a meridie usque ad 8, a sore fit.
3 Nov.	0-£ +0 a meridie hora 4.
4 Nov.	I was directed to my voyage by the Earl of Leicester and Mr secretary Walsingham, hor9.
7 Nov.	I came to Gravesend.
9 Nov.	Iter transmarinum. I went from Lee to sea.
14 Nov.	I came to Hamburg: hor. 3.
6 Dec.	I came to Berlin.
11 dec.	To Frankfurt-upon-Oder.
15 Dec.	News of Turniser's coming, hor. 8 mane by a special messenger.

Dee had shortly afterwards recorded in his special notebook, the 'Compendious Rehearsal,' the following:

My very painful and dangerous winter journey (about a thousand five hundred miles by sea and land) was undertaken and performed to consult with the learned physicians and philosophers beyond the seas for her Ma(tie's) health-recovering and preserving.

Dee had been sent across the channel to the Continent to consult with physicians there regarding the Queen's health, which was endangered by some unnamed malady. He was told to return within a hundred days, i.e., at the beginning of February, 1579. Dee was back in the country by March, 1579, for his diary records "A moist March and not windy." His hundred days had been hard spent.

The entry for 15 December was heralding the meeting that had been arranged between Dee and "Turniser" while Dee was in Germany. This Turniser's full name was Leonard Thurneysser, who had been the personal physician to the Elector John George of Brandenburg in Frankfurt. His wife, Sabina of Brandenburg had been 'healed' by Thurneysser, who was then considered a 'miracle doctor.'

The nature of Elizabeth's illness is not for discussion here. I would, however, wish to comment on her level of trust in Dee. It was clearly 'absolute,' as must also have been the trust placed in him by those intimately associated with her, especially the faithful and ever-close Robert Dudley, Earl of Leicester, and Sir Francis Walsingham, her security man. Jane Fromonds, Dee's newlywed young wife had visited the Queen herself on 22 October before Dee's epic journey.

Dee's mission is highly significant. His diaries, which include symbolic coding, show him to be a medical man of no mean stature, who recorded with discretion his studied observations of the monthly cyclical patterns of women close to him, including their pregnancies. This was quite apart from other aspects of his expertise. Although the Queen lived for another twenty-four years, her state of health at this time was a matter of real concern and persuaded Robert Dudley, her near-constant companion, intimate, and official servant to the Queen, to personally accompany Dee to the point of departure on this long and arduous journey.

Blanche Parry would have been well-aware of her mistress's condition. As godmother to Dee's first-born son Arthur, she was also in a totally unique position of trust. It seemed to me that Queen Elizabeth and her ancestors were able to depend on the loyalties of families with connections to this beautiful area of Tintern, which was separated from England by the river Wye. The Nurtons, particularly, had a special role in these proceedings, but first we must look at an important aspect of Shakespeare's legacies.

Shakespeare's Sonnets were first published by Thomas Thorpe and were entered in the Stationer's Register on 20 March 1609 shortly after Doctor Dee's death in 1608. The volume consists of one hundred and fifty-four separate groupings of verse representing different aspects of human patterns of behaviour, experience and relationships.

I include Sonnet number '72,' a number central to the geometry of Byrom's collection of drawings in its entirety:

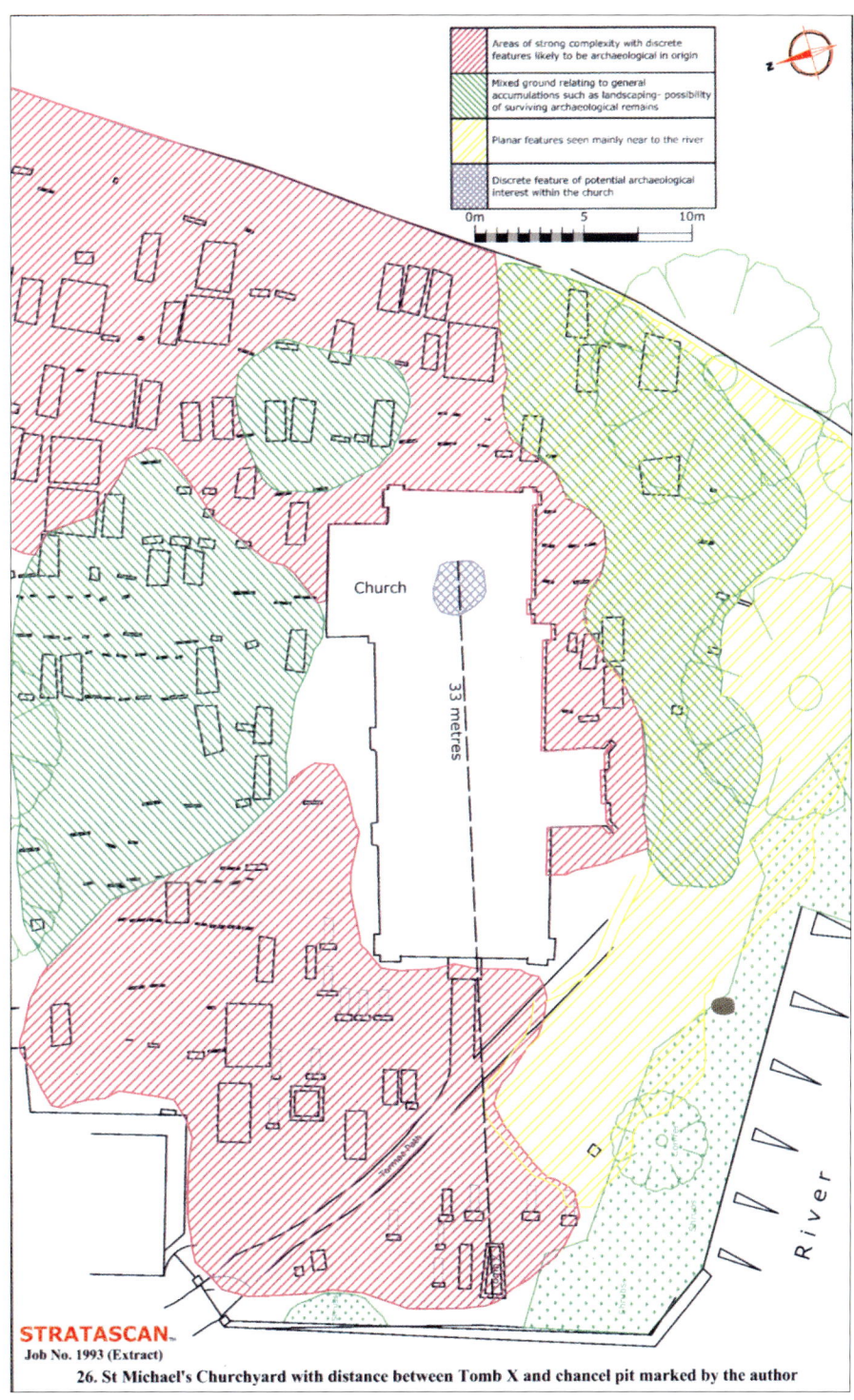

Church

33 metres

River

STRATASCAN.
Job No. 1993 (Extract)

26. St Michael's Churchyard with distance between Tomb X and chancel pit marked by the author

The tomb of interest. Measurement

LXXII

O lest the world should task you to recite
What merit liv'd in me, that you should love,
After my death, dear love, forget me quite
For you in me can nothing worthy prove;
Unless you would devise some virtuous lie,
To do more for me than mine own desert,
And hang more praise upon deceased I
Than niggard truth would willingly impart:
O, lest your true love may seem false in this,
That you for love speak well of me untrue,
My name be buried where my body is,
And live no more to shame me nor you.
For I am sham'd by that which I bring forth,
And so should you, to love things nothing
Worth.

Apart from the significance of its place as number seventy-two in the volume, the piece focuses on the doings of an author who considers himself to have led a life of lies and deceit. Apart from the tragedy implicit within the verse is the eleventh line:

My name be buried where my body is

Of course, he does not say where that will be. Whom he is addressing as "dear love" is another matter. So, I returned yet again to the site which has become as a 'magnet' to me – St. Michael's Tintern, looking for more clues and especially answers. Doctor Dee had been 'told' to be back from his "painful and dangerous journey within a hundred days." What was so significant about a hundred days? Number symbolism was very important to the Elizabethan elite, as we know. It was the tombstone in the graveyard at St. Michael's Church, Tintern, with the four letters 'C' on the sword engraved at the top which attracted my attention. 'C' stands for 'Centum,' the Roman word for one hundred. In related number symbolism 100 can be identified with Francis Bacon's name, a matter I will focus on later. It was the Queen who directed Dee's return date. Was it a coded signal between the two of them?

The findings of the Stratascan Radar Report that I commissioned at St. Michael's demonstrates that the distance from the edge of the tombstone to the chancel pit of the church itself is 72 cubits (33 metres). The number 72 represents the numerical equivalent of the Greek letters for the word 'truth.' In fact, there are biblical, cabalistic as well as astronomical associations with the number 72. I believe that the positioning of the gravestone in relation to the chancel pit is deliberate.

My thoughts regarding these findings have already been outlined in "The Hidden Chapter," where I explained the connection of Bacon, Dee, the Sidney family, the Beaufort family and Walsingham to the brass-works nearby. I now had enough clues to repeat the suggestion that there was a significant burial in the chancel of the church. I think it was the final resting place of Francis Bacon. The stone in the graveyard serves as a symbolic clue.

The Patron of St. Michael's church at the time of Francis Bacon's death, as recorded on 9th April 1626, was William Herbert, 3rd Earl of Pembroke. I have already noted that he and his brother, the 4th Earl of Pembroke, were the dedicatees of the first publication of Shakespeare's folio of 36 plays in 1623.

The "Sonnets" were first published in 1609. At the front of the publication is another intriguing dedication:

> To the Onlie Begetter of
> These ensuing Sonnets
> Mr. W.H. all happiness.
> And that Eternitie
> Promised
> By
> Our ever-living Poet
> Wisheth
> The well-wishing
> Adventurer in
> Setting
> Forth
> T.T

There has always been a considerable amount of speculation as to the actual identity of "Mr. W.H." The most popular candidate for this recorded accolade was the 3rd Earl of Pembroke, William Herbert. I would agree, but my opinion has been influenced by 'Sonnet 72' as well as other associations with St. Michael's church in Tintern and The Nurtons. My suspicions at the time caused me serious concern. The author of 'Sonnet 72,' and all other sonnets in the publication have been attributed to William Shakespeare, who's burial memorial is in The Church of the Holy Trinity, Stratford-upon-Avon.

The Patronage of St. Michael's church had to be looked at even more closely, especially during the years that the 3rd Earl of Pembroke (Mr W.H.) held certain responsibilities in connection with the administration of the church as well as the site itself. Briefly returning to the 'Sonnet' dedication by Thomas Thorpe, there is a suggestion in it that 'Mr. W.H' was to be involved in sea-faring ventures, since travel by land would not be so noted. The 3ʳᵈ Earl of Pembroke William Herbert was known to be very interested in the planned expedition to 'New England' and became a member of King James's council for the Virginia Company of London in May, 1609. There has also been some doubt as to the correctness of this Pembroke's attribution as the dedicatee, because of the reference to 'Mr. W.H.' rather than to his 'Lordship' or 'Earl.' William Herbert was born on 8 April 1580. He inherited his title in 1601. As a godson of Queen Elizabeth he was a popular figure in court circles, single-minded, though of a melancholy disposition in his youth.

It was his death on 10 April 1630 which aroused my curiosity even more. Stored in Cardiff Library in Wales is a manuscript entitled 'Herbertorium Prosapia.²' The manuscript is a history of the Herberts and was compiled by Sir Thomas Herbert of Tintern (1606-1681). He had been the owner of The Nurtons from 1640 until his death. His patron in early adulthood had been his kinsman this same William Herbert, the 3ʳᵈ Earl of Pembroke. His respect and devotion to the Earl can be seen and understood in the entry I choose to quote here from the Cardiff Manuscript in its entirety. It follows the death of Henry Herbert the 2ⁿᵈ Earl of Pembroke in 1601:

William Herbert his eldest son succeeded in the Earldom of Pembroke, a noble man of extraordinary worth generally beloved for his integrity, learning, piety, vertuo gentleness, courage and prudence in affairs of state and exceedingly beloved by King James and King Charles his son (at K James coronacon 1603 (sic) he bare the sword as his Ancestors and other Earls of Pem had done). He was a privy councellor to both, made Knight companion of the noble order of the Garter and invested with the habits and Ensigns thereof in the 1. yeare of King James. In the 7. yeare of his Reign he was made Govenour of Portsmouth, in the 15. yeare he was chosen chancellour of the University of Oxon and next yeare Lord Chamberlain of his maj household, Gentleman of his maj Bedchamber, Justice in Eyre of his maj forests and chases South of Trent, Lord Warden of the Stannerhs in Devon and Cornwall, Lord Steward of his majs house and at the coronacon of King Charles which was in the yeare 1625 was joined in commission with Thomas Howard, Earl of Arundell to make 'Knights of the Bath', as the King should then call to that dignity to attend at his Coronacon, the first of them being St Philip Herberts son, the Lord Charles Herbert and was mecenas to men of parts a true and reall upholder of learned Endevours as it is said in the dedicatory Epistle to Epiclolul.

This William, Lord Herbert Earl of Pembroke married the Lady Mary Talbot eldest daughter of Gilbert Lord Talbot the 7 Earl of Shrewsbury of that name by which lady he had one son called Henry who died in his infancy in the yeare of our Lord 1621 to the great grief of Parents especially of the mother while she lived.

He had two natural Children by the Lady Mary Wroth, the Earl of Leicester's daughter Wm who was a captain under Sr Hen. Herbert Collonell under Graiw Maurice and dyed unmarried and Catherine the wife of Mr Loud neare Oxford.

This noble Lord the night before he dyed supped at Fishers Folly without Aldgate with the Lady Lucy Countess of Bedford, Countess of Devons Mrs Murray with severall other Lords and Ladyes. And some of the company – discourcing upon the vanity of this life, the certainty of death and the uncertainty of the hour of death; The Earl tould the company that God Almighty had kept secrett the end of mans life, that he might be prepared – and in continuale expectacon of it. But says he would have norw trust in the presumptons of Judiciall Astrology, for I bless God that I am now in as good health as ever I was in my life and have no simptoms of death or other malady; nevertheless my Fathers Chaplain, being one very Studious in natural Philosophy at my Birth calculated the horoscope of my nativity, and found that I should end my life this day: The Company agreed with the Earl as to the fallaciousness of that study, and a little before midnight parted; But being come to Baynards castle and in bed, he began to complain, and growing worse the Countess rung a silver watch or bell, which being heard, the servants hastened to know the occasion, and found the Earl dead of an Imposthumo or Apoplexy: Lamentable were the cryes of his good lady and family and no less the sorrow of all that knew him. I heard our gracious King Charles the first say, when the newes was brought him of this good Earls death, he was never more surprised nor fuller of anguish and sorrow for the death of any of his servants or Councellors than for him, for he was a father to him in affection – faithful and prudent to him in Advice, an honour to his Court and family – not sparing his own Estate, which was great, £30000. P anno as I have heard Mr Halseworth his secretary affirm to honour the King, and never begging of anything of the King, but Sr Gervace Elvis his Estate, which was considerable, being executed when he was Lieutenant of the Tower for being accessory to Sr Thomas Ovenburys death. (who advised the Lord Carr, Earl of Somerset not to marry the Earl of Essex his repudiated wife, which the King readily granting the Earl and freely gave it his widow the same day for the relief of her and her children, which when the King heard he highly praised the Earl for his care and kindness to the innocent). He was of the ancient stock of nobility, very comely in person, a gallant courtier, a great lover of the King and Royal Family, a friend to good men, an enemy of vice, charitable to the poor, liberall to men of desent, affable, yet grave and wise in councell, he was from Baynards castle where he lay for some days (from the 10th of April 1630 when he departed this life conveyed to Wilton, very many coachess of nobility attending the Corps, whence he was carried to Salisbury with all honorable and due solemnity and interred in the Cathedrall neare his father; he dyed in the fiftieth yeare of his age, leaving his brother Philip his heire who succeeded also in his Dignity.

We must be reminded that this anecdotal quote from the Herbert miscellaneous notebook was put together by a beneficiary of the 3rd Earl's Patronage, Sir Thomas Herbert of Tintern. The copy I used contained spellings not easy to read. I reproduce my copy as understood by myself at the time. Although written in retrospect maybe years after the event, it is clear that the reflections contain knowledge of 'in-house' intimate information.

The Third Earl William Herbert died suddenly and unexpectedly, having enjoyed a celebratory dinner on 9 April. His actual 50th birthday had been the day before on 8 April. The venue was 'Fisher's Folly' in London. It was the date which aroused more than a little interest. It was the same day of the year as Francis Bacon's own recorded death a few years earlier on 9 April 1626. Fisher's Folly was owned by Francis Manners, 6th Earl of Rutland. His brother Roger, 5th Earl of Rutland married Philip Sidney's daughter, Elizabeth. The two Manners, as Earls of Rutland, become important later in the story.

I began to consider whether the Earl's death in 1630 of a possible 'apoplexy' was a convenient diagnosis. Was there a more dubious explanation? In Thomas Herbert's account he does not give a complete list of those present at the celebratory dinner. However, it seems that the 3rd Earl's food platter, together with what he had to drink may well have contributed to his fatal collapse so soon after he arrived home. His predicted destiny had certainly taken place, although in discussion that possibility was disputed at the table.

Without knowing exactly the names of all who were present at the dinner one must be very careful, from such distance in time, to apportion suspicion or a likely explanation for what appears to be an untimely event. Thomas Herbert certainly presents the 3rd Earl as an exemplary character with fine attributes, presentable and very popular.

However, data from the period of the 3rd Earl's life does show that around the time of his succession to his father's estate in 1601, he was committed to Fleet Prison in London for refusing to marry Mary Fitton who had become pregnant by him and whose child subsequently died soon after its birth. Herbert's stay in prison was short. Mary Fitton was a member of the Holcroft family, a family of some stature in its own right, who by marriage was closely connected to the Earls of Rutland, the owners of Fisher's Folly, the Manners brothers.

For the historian and researcher there are sometimes problems connecting families with each other in that married women usually took the names of their husbands. Titles may camouflage complex family alliances, in many cases unwittingly so. The 3rd Earl, William Herbert, would have been barely twenty-one years of age at his considered act of indiscretion in 1601. Could deep family memories bear a grudge for the next thirty years? The Earls of Rutland, the Manners brothers, have been for me a family full of historic intrigue.

But there are other facets of William Herbert's life which became a focus for me to consider. The authoritative Burke's Peerage states that William Herbert became "Grandmaster of Freemasonry in 1618," a position he held until his death. The title of Grandmaster was that of the head of the organisation. Modern Freemasonry is said historically to be a society with secrets rather than a secret society. The first Grand Lodge of Freemasonry was inaugurated in 1717 in London. Therefore, an organisation with a title identical to that, prior to the date of founding the Grand Lodge suggests an 'order' different in some way, but related. The 3rd Earl was the figurehead of it for some years. The ambitions and legacies of the organization are unclear. The Earl took over the position from Inigo Jones, the prestigious Renaissance architect. This fact persuaded me to consider whether Herbert's title was connected to a 'medieval guild,' the "London Company of Masons and Freemasons," which itself had been in existence since the fourteenth century. Various modern guild historians have assessed their records and I quote from two of them. Carew Hazlitt writes of the fourteenth century members:

> Freemasons represent in a modified form the Society which once rose to exceptional eminence and acquired even formidable power. They enjoy the unique distinction of having laid the basis of a social cult which has immeasurably outstripped its founders.

Still another of the guild historians, Robert J. Blackham, says that the:

> Masons Company...has not only preserved its own identity from the Middle Ages, but played a considerable part in the early development of a world-wide and esoteric organisation which has adopted an ancient title... The London Livery Company of Masons may proudly claim that it is the principal connecting link in the chain of evidence which indicates that the modern social cult known as "Free and Accepted Masons" is lineally descended from the old fraternity of masons which, in association with the monastic orders, built the stately Gothic buildings of the Middle Ages.

Carew Hazlitt and Robert Blackham's quotes are from the book "Freemason's Guide and Compendium" by Bernard Jones[3]. Although the Peerage Records do not actually state that William Herbert's status as 'Grandmaster' is of that specific organisation, he was involved with a society with secrets similar. In addition, the Earl was Patron of St. Michael's church in Tintern at a time when the history of the site shows it to be steeped in a 'social cult' going back to at least the Roman period.

The Reverend Neville Barker-Cryer had identified and recorded several exclusive Masonic features within the boundaries of the graveyard and Roman pottery was discovered during my commissioned archaeological project.

But it was Sir Thomas Herbert of Tintern, the collator/author of the Cardiff manuscript, who provided the biggest mystery surrounding the 3rd Earl of Pembroke. He published a book in 1677, four years before his own death, entitled "Some Years Travels into Divers Parts of Africa and Asia the Great."[4]

Chapter one of the book is entitled 'Travels – Begun Anno 1626.' The first line of the chapter begins:

> Upon Good Friday, in the year 1626, we took at Deal near Dover, having six great and well manned ships in company all which were bound for the East Indies.

Sir Thomas is writing of travels he embarked on beginning on that date. The 3rd Earl of Pembroke, William Herbert, was his Patron, and holding significant sea-faring appointments, had arranged for him to join the 'suite' of Sir Dodmore Cotton, the ambassador to the King of Persia on journeys which were to last for three years. The ship was called 'The Rose.' Sir Dodmore Cotton was the son of Sir Robert Cotton, the antiquarian who discovered the original diaries of John Dee, having dug them up in a field very close to Dee's house in 'Mortlake.' Coincidentally the date of this epic journey began on Good Friday 1626. Why the coincidence?

In the account given by William Rawley, Francis Bacon's personal secretary and chaplain, he records the following entry of Bacon's death:

> He died on the ninth day of April in the year 1626, in the early morning of the day then celebrated for our Saviour's Resurrection in the sixty-sixth year of his age.

Good Friday, 1626 would have been the 7th April. However, the date in the Christian Calendar of Jesus, the Saviour's resurrection would have been 9th April (Easter Sunday). After registering the departure date of Good Friday Thomas Herbert goes on to say:

> In a few hours coasting close by the Isle of Wight (called so from Gwydir, a British word, signifying cut off or seen at a distance) a sudden borasque or gust assaulted us; which after an hours rage spent itself, and blew us the third day (double solemnised that year by being the Feast of Mother and Son) upon the Lizard's Point, the utmost promontory of Cornwall as we passed.

It would seem the 3rd Earl had been privy to and part of events which included yet again the date '9 April,' the date of his own fateful dinner at Fishers Folly four years later. It also follows that Thomas Herbert chose to record the third day's event of this epic journey with "double solemnised that year by being the Feast of Mother and Son." That must have been the Easter Sunday after Good Friday.

Both Thomas Herbert and William Rawley had specifically highlighted a Christian event, 9 April Easter Sunday, 1626 in their notations. the first in connection with a sea voyage and the other marking the death of Francis Bacon. There seemed to me some sort of overriding coincidence here. The 3rd Earl of Pembroke seemed to be 'hovering' over the two events and dates, especially since the same date would play such a significant role in his own demise. At this point I am beginning to suspect that there may be more than mere coincidence regarding the date 9 April, especially considering that the date becomes a very significant one recurring throughout my investigations. Is it coincidence? Providence? What is that saying. The luck of the draw? Somehow, I think not. Somehow, more is involved than meets the eye. I am simply the investigator noting events and making observations. John Dee could perhaps advise. "Curiouser and curiouser," someone said. I agree.

Thomas Herbert had acquired so many diplomatic roles in his life, requiring him to be dependable and reliable, so any clues on his part would be laid down with absolute discretion. Being part of the 'suite' of Sir Dodmore Cotton carried with it discretionary responsibilities. The arrangement had been set up by the 3rd Earl W.H. and family anecdotes would be exchanged at some point during the trip. More than likely Dee's original manuscripts would be a topic for lively conversation given the fact that they were found buried in a field by Dodmore's father, and then there was Dee's established reputation. By the time Thomas Herbert published his account of the voyage fifty years had come and gone. He had been a witness to other unique historic events, not least of which was the execution of King Charles 1st. I had come to regard Thomas Herbert as a very reliable and discrete chronicler.

Having noted carefully the details surrounding 9 April 1626, I looked at the beginning of Herbert's book for any other meaningful facts and recognised another interesting observation which he chose to record so many years after the event. It must have made an impression on him at the time. He not only gives a Catholic Christian label to the third day of the voyage but states that they were:

Upon the Lizard's Point the utmost promontory of Cornwall as we passed:

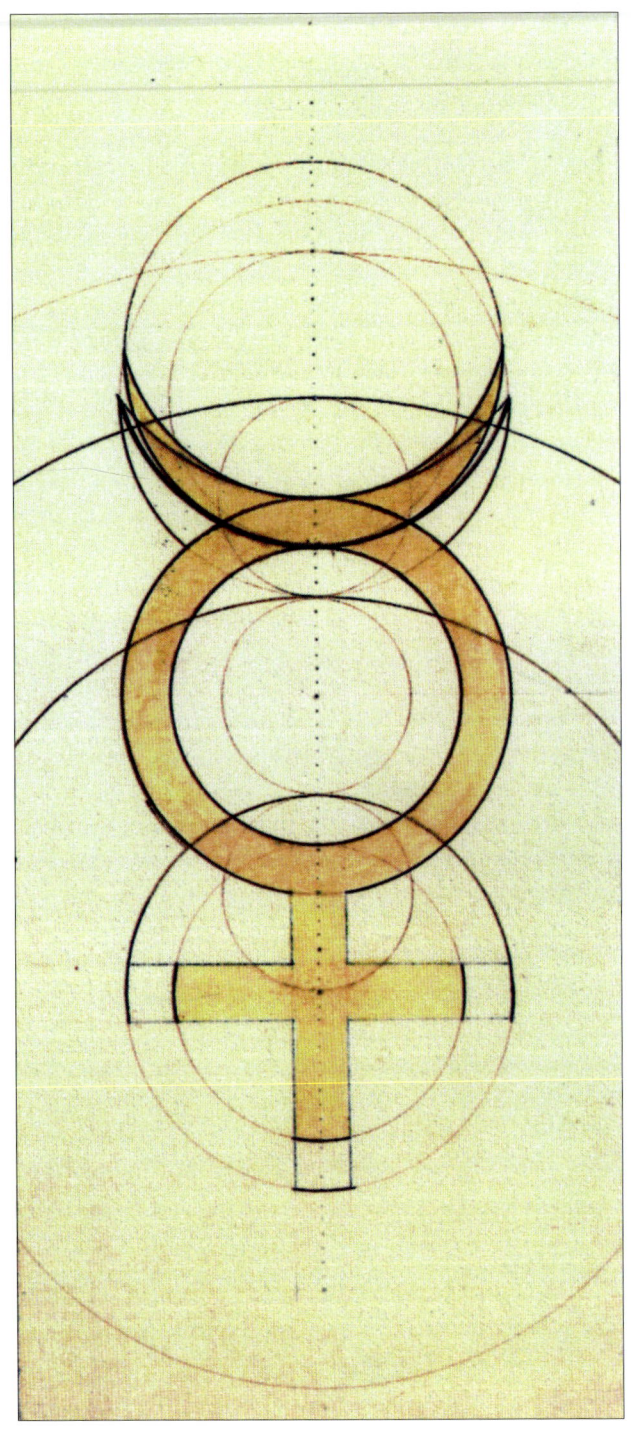

The Monad. John Dee

Lizard's Point would certainly have made an impression on him as it has for all recording historians ever since. On the 29th July 1588 at three o'clock in the afternoon the first sighting of the Spanish Armada was witnessed from Lizard's Point – one hundred and twenty ships with twenty-nine thousand men on board. The 3rd Earl of Pembroke would have been eight at the time, but for Queen Elizabeth 1st it was one of the most memorable events of her reign. The occasion was a failed attempt to restore the Catholic faith to a Protestant Britain. In a noble act, the Queen went to address her troops at a critical juncture, accompanied by her devoted companion, Robert Dudley, the Earl of Leicester. He walked alongside her horse, bare-headed, and died a few weeks later on 4 September of an uncertain illness. His death was also unexpected. The Queen was deeply affected and remained confined to her quarters for some time. She referred privately to Robert as her 'eyes' and kept his last letter to her in a box beside her bedside until she herself died some fifteen years later. It was known by those close to Queen Elizabeth just how devoted she and Robert were. I suspect that Thomas Herbert acquired confidences years later from William Herbert, 3rd Earl of Pembroke.

It is clear that Thomas Herbert was deeply affected by the Third Earl, (his patron's) death in 1630, having only returned from his long voyage in 1629. By 1640, we learn that Thomas Herbert became the owner of The Nurtons in Tintern. By that time, it was already the 'Hermetic College' founded by yet another William Herbert. This William Herbert was an intellectual who considered himself to be a descendant of the 1st Earl of Pembroke, Sir William Herbert, who died in 1469 and is buried in Tintern Abbey. The Nurtons 'college buildings' were situated very close to St. Michael's Church.

What is of real significance is the loyalty to Queen Elizabeth by these families. William Herbert had an even closer relationship with John Dee. There is an entry on 1 May (1577) as follows:

> I received Mr William Herbert of St. Gillian [St. Julian's] his notes upon my Monas.

Dee is referring to his classic text, "Monas Hieroglyphica," which Dee himself confidently describes as a book of "great rarity and remarkable quality." It was first published in 1564. A drawing of the symbol for Mercury, the Messenger, is on the Frontispiece of the book and is part of the sequence of the Globe theatre illustrations in the Byrom collection. So who was this "Mr. William Herbert of St. Gillian" (sic)? John Dee clarifies his identity for us with a mysterious entry in his diary some two years later. On 31 May 1579 in Greek symbols he notes that Sir William Herbert of St. Julians visited him and went on towards London. It is generally accepted that Dee entered coded symbolism when the entry was to be regarded as secret or personal.

This Sir William's association with Dee developed to such an extent that entries show that he acquired a house in East Sheen, close to Dee's residence at Mortlake. I was also reminded that 'Mr. John Herbert of East Sheen' became the 'stand-in' godfather at the christening of Arthur Dee. Their children Arthur Dee and Mary Herbert are described as playing together in an entry for January, 1582. So by this time, their corporate family and scholastic links had already lasted for five years.

Sir William (not the Earl) had been knighted by Queen Elizabeth at Richmond on 21 December 1578 whilst Dee was abroad on his special medical mission on the Queen's behalf. The secretive 1579 meeting at the end of May is a few weeks before his son Arthur's christening, as already noted. Sir William Herbert's principal estate, St. Julian's, was situated not far from Newport in Monmouthshire, not far from Tintern. He had inherited the property and land in Tintern Parva, close to St. Michael's Church, now referred to as The Nurtons. By 1585, it is known that Sir William owned part-shares in the Tintern Wire Works, otherwise known as the Brass Works. For the next three years he was in Southern Ireland developing colonies on behalf of the Crown. He returned in the spring of 1589 and by September of that year he was resident at his Tintern Parva property. It would be during this time that his Hermetic College was established. Unfortunately, on 4 March 1593 he died at the approximate age of forty.

Sir William had been married to Florencia Morgan and they had three children – two sons and a daughter Mary. The two small sons had died tragically, having eaten some candy containing rat poison planted in their father's library in an effort to eliminate the visiting rats. They had come upon the candy themselves and eaten it. They could not be saved. The parents were distraught and never fully recovered. William Herbert, now with daughter Mary as heiress to his considerable estate, left an instruction in his will that she should marry a member of the Herbert Family dynasty; she subsequently married a kinsman, Edward Baron Herbert of Cherbury in 1598. He was the brother of the poet George Herbert. But significantly, there was another brother, Sir Henry Herbert. Sir Henry was the Theatre Censor of plays in England from 1623-1641 when all theatres were closed during the Civil War. He retained the position despite the closures until the Restoration, when he became 'Master of Revels' until his own death in 1673. His influence therefore was the 'power' behind public performances, including the Globe. In addition, Mary Herbert and her husband Edward were responsible for her father's legacy of the Hermetic College. Shakespeare's plays would inevitably require official permission to be acted and be under review before presentation. It was Mary and Edward's son Richard, 2nd Lord Herbert of Cherbury from whom Thomas Herbert (the chronicler) acquired the property and all that it meant in 1640.

The nearby St. Michael's Church was under the patronage by this time of Philip, 4th Earl of Pembroke, the brother of the 3rd Earl. These brothers were the dedicatees of the first folio of Shakespeare's plays, which had been published in 1623, the same year that Mary's brother-in-law became Censor. The power and influence over the cultural heritage of the country seemed to be under the control of the Herbert/Pembroke dynasty, brokered through designated groups and property holdings, the Hermetic College being one of them.

John Dee's influence had been enormous in the formative years of the College. Dee considered architecture the highest among the arts and sciences and in his Mathematical Preface to "The Elements of Geometry of Euclid," which had been translated by Henry Billingsley and published in 1570, he summed up Vitruvius's view of architecture: [5]

"An Architect (saith he) ought to understand Languages, to be skilful of Painting, well-instructed in Geometrie, not ignorant of Perspective, furnished with Arithmetike, have knowledge of many histories, and diligently have heard Philosophers, have skill of Musike, not ignorant of Physike, know the aunsweres of Lawyers, and have Astronomie, and the courses Caelestiall in good knowledge."

What is highly significant here is that Henry Billingsley the translator of this "Elements of Geometry," lived within a few miles of Tintern. He rented and then mortgaged properties from Sir William Herbert of St. Julian's for his own use. Entries in Dee's diaries confirm that Sir William Herbert, founder of the College, discussed Dee's masterpiece entitled "The Sacred Symbol of Oneness" with him in 1577. The principle symbol of this thesis is the Pythagorean 'Y.' The philosophy within the book describes the strengths and weaknesses of the human condition and life itself as related to nature, numbers and Pythagorean philosophy. (see p. 244)

As already discussed in my earlier books, I had numerous meetings with Elsa and Adrian Wood, the current owners of The Nurtons, where discoveries and ideas were exchanged. They agreed that the site had at one time housed the Hermetic College. I had discussed with them, particularly, John Dee, Francis Bacon, as well as Thomas Herbert, the chronicler of so much that I had discovered. They had given me access to property research papers of their own. Sometimes fresh insights occur unexpectedly, so it was in 2009, the very year that I had my last contact from Lord Michael Birkett, Elsa and Adrian told me of a new discovery at The Nurtons, which was of great interest to me. I include their commentary of the event. It seems appropriate at this juncture:[6]

The Nurtons, Pythagorean 'Y'

"The Nurtons"
'Y'

In 2009 we renovated one of our outbuildings for accommodation. In order to install underfloor heating, we lifted the old flagstones from the floor. On the undersides of about 12 of these stones we noticed the letter Y had been carefully carved into the surface of the stones. These were not superficial scratches as in many masons' marks we have seen previously but were well indented in the stones to a depth of 3-4 cm. This must have taken some time and the letters had been done neatly and deliberately with lines at the foot and over the top of the arms of the Y. Alongside some of the Y's there were Roman numerals carved to a similar depth. These numerals were also on stones where there were no Y's -the numbers did not exceed VIII and II – VIII are repeated on different stones. (If this were a numbering system why no more than VIII?) Unlike the Y's there were no lines on the numeral at top or bottom as is conventional for Roman numerals.

It is possible that the stones had already been removed from somewhere else and re-laid where we found them. Several of the stones had been fractured and excavations over the years have revealed substantial and well-constructed stone walls below ground surface on the site indicating there were earlier buildings on the site, now demolished.

Most of the flags have been used to lay the patio in the back yard with the Y's downwards to prevent them being worn away but 2 have been retained to be visible as display stones. (The position of these stones has been mapped.)

Other flagstone floors exist at the Nurtons that have not been disturbed.
Elsa and Adrian Wood

This is the same site where William Herbert had set up his Hermetic College. Sometimes truths reveal themselves at unlikely times. For me, Elsa and Adrian's finds were reassuring and timely. At this point I reminded myself of another fact. Lady Anne Bacon, Francis Bacon's reported mother, had spent her last years at Mount St. Albans situated practically next door to St. Julian's, Sir William Herbert's manor house about twelve miles from Tintern. She died there in 1610. At that time, William Herbert's daughter Mary and her husband Edward Herbert would have been her neighbours. I quote the contents of a letter written by Francis Bacon to his kinsman Sir Michael Hickes on 27 August 1610 regarding Lady Anne's funeral: [7]

It is but a wish and not any ways to desire it to your trouble. But I heartily wish I had your company here at my mother's funeral, which I purpose on Thursday next in the forenoon. I dare promise you a good sermon to be made by Mr. Fenton, the preacher of Gray's Inn; for he never maketh other. Feast I make none. But if I mought have your company for two or three days

at my house I should pass over this mournful occasion with more comfort. If your son had continued at St. Julian's it mought have been adamant to have drawn you; but now if you come I must say it is only for my sake.
Yours ever a friend,
Fr Bacon'

There is no doubt that the letter is authentic. I have a photograph of the original from the British Library Lansdowne Collection. Francis Bacon, in referring to Sir Michael's son, is writing of his fourteen years old son, also named Michael. It is clear he had been staying at St. Julian's. Henry Herbert, the future Master of Revels would have been fifteen years old at the time and a brother-in-law of Mary Herbert. It is more than likely that he was at St. Julian's at the time and the two teenage boys were friends.

It is my opinion that Francis Bacon's mother, Lady Anne, and years later Sir Francis himself, were buried in St. Michael's church in Tintern Parva by arrangement with Mary Herbert and her husband Edward, together with the patrons of the church, the 3rd and 4th Earls of Pembroke.

For me, the curtain had been moved a little further aside.

CHAPTER TEN

Disturbing Facts. Collected

THE CLOSE AND secretive connections recorded in John Dee's diary regarding his meeting with William Herbert of St. Julian's and his long-term association with Blanche Parry, Queen Elizabeth's confidant, were enough to persuade me to look for additional data. The 'Pythagorean Y' stone slabs at The Nurtons had become something of an enigmatic clue to possible other pieces of the puzzle. The slabs were revealed in 2009.

It was after spending some time with yet more catalogues and indexes in libraries, book shops and on the internet that I came to learn of yet another edition of John Dee's diary, edited by John Eglington Bailey in 1880, an antiquarian who had worked for Ralli Brothers (a Greek business family) in Manchester for most of his working life. Bailey's version concentrated on Dr. Dee's years in Manchester during the time when he was the appointed Warden of the Collegiate Church, now the Cathedral. The years covered were 1595-1604. Bailey used as his source material John Dee's original manuscripts at the Bodleian Library, Oxford. I acquired a copy of his book. Having spent many years studying Fenton's edition of Dee's diaries, I noticed a significant difference in the last entry of Bailey's printed version. I record that 1601 entry by Dee in full:[1]

> April 9 (Monday), Mr. Holcroft of the Vale Royal his first acquaintance at Manchester by reason of Mr. William Herbert his servt (not friend), etc. He used me and reported of me very freely and wurshipley etc. Mr Holcroft of Vale Royall his most friendly acquaintance and using of me.

The Fenton version's entry by Dee reads as follows:[2]

> 27 Mar. Mr Holcroft of Vale Royal his most friendly acquaintance and using of me. Mr Holcroft of Vale Royal his first acquaintance at Manchester by reason of William Herbert etc. He used me and reported of me very friendly and worshiply.

While the gist of the entries is the same, the dates are different. Bailey's Dee entry date is '9 April,' that significant date again. The Mr. William Herbert referred to has to be William Herbert, 3rd Earl of Pembroke, who succeeded his father to the title the same year. 'Mr. Holcroft of Vale Royal' was Thomas Holcroft (1557 – 1620). His first wife was Elizabeth Fitton (died 1595). She was the sister of Mary Fitton who had a child by this William Herbert, noted earlier. Many years later William Herbert celebrated his fiftieth birthday on that same anniversary date at Fisher's Folly and died in the early hours of the following morning, as previously noted. I found this entry of great interest, especially since it was the last recorded entry in both diaries of Dee.

But why had the editors Fenton and Bailey recorded the same content on different days? Bailey provides the necessary clue. His Dee entry for '19 March Thursday' 1601 includes:

9th anni veteris

When translated this means 'calendar – old style.' Fenton's entry is dated 9 March. He omits the Latin quotation noting the calendar change. Editor Bailey's recorded Dee entry for 19 March is as follows:

I receyved the long letters from Bar Hikman hor 2 a meridie by a carrier of Oldham.

Fenton's content of the entry is the same. It becomes clear that Dee himself in his original manuscript uses the updated calendar (ten days difference) for sections of his diary, which Bailey honoured accurately as the date recorded in the diary. Fenton on the other hand decided to use the 'old style,' thinking it would be helpful to the reader. But that would be misleading. In this instance, Bailey's representation of the original entry, accurate as Dee recorded it, was a reminder of the significance of 9 April. Dee recorded no more daily doings in his diary after that, which seems a fitting end to the record.

It seems that Theodore Dee, Dee's son aged thirteen years, died on that day or thereabouts. In the Burial Record of the nearby Collegiate Church of which John Dee was Warden, there was an entry for 12 April, reading as follows:

Tidder former son of John Dee Esquire. Warden

Further up the page in the original Collegiate records is the entry:

Burials in April 1602

It seems that there could be difficulties understanding the church records. Most chroniclers report that the son died around the time of Dee's last diary entry in 1601. In his diary for 2 December (1600) there is another intriguing entry:

> College audit. Allowed my due of 7li yearly for my house – rent till Michaelmas last. Arthur Dee a grant of the Chapter clerkship from Owen Hodges, to be had if 6li were paid to him for his patent.

In the event, it seems that Dee's son Arthur as 'Chapter Clerk,' a young man by now, was responsible for burial records at that time. The confusion surrounding the entry of Arthur's brother Tidder Dee's burial is worthy of note. 'Tidder Dee' was christened Theodorus Dee on 1 March 1588, having been born the day before. The family was living in Trebon, part of the Czech Republic at the time (a six year stay). They arrived back in England on 22 December 1589 and two weeks later John Dee was visiting Queen Elizabeth I at Richmond Palace. Not much time wasted for that meeting then!

Dee had left England in September, 1583 accompanied by an entourage of family and consultant staff, including the scryer ('seer') Edward Kelly with a library of learned books and manuscripts. Dee's scientific skills and philosophical and medical knowledge were much in demand by European aristocrats. His return to England after a six-year stay on the continent was not under the happiest of circumstances. The situation surrounding Theodore's birth and arrival in the world was partly responsible, a matter to be discussed more fully later.

Returning to Dee's last entry in his diary and the mystery surrounding 'Tidder's' death, the entry became even more of a puzzle, especially because other nearby entries for previous weeks and months were few and far between. The one before the 9 April entry (Monday) records the arrival of Mr. Holcroft on Sunday. Holcroft, for many reasons, had no love lost for Francis Bacon or anything to do with him, a matter we will return to shortly. The visit remains somewhat mysterious in nature, Holcroft being a member of a family close to Robert Devereux, the 2nd Earl of Essex, who had been executed for treason. The discussion must have included the circumstances of the trial and outcome, although Dee does not record it in his diary. Amongst this activity, a row occurred in Dee's household centred round his son Arthur, the young man. It was the entry before that visit from Holcroft that notes the old calendar usage and the receipt of letters from 'Bar Hikman' on Thursday 'Mar 19.' A previous 2 March (Monday) entry records 'Mr. Roger Koke went towards London.' But it was the entry before that which gave fresh impetus to my investigation. It reads:

Feb. 25 (Wednesday) R.K. (oke) pactum sacrum hor. 8 mane.

Translated from Latin it means a 'secret and sacred understanding' was being recorded. Koke was also a noted 'scryer' who knew John Dee intimately. The date and time of day of the entry were the same as the date and time of day when Robert Devereux, the 2nd Earl of Essex was executed at the Tower of London. Was this entry made to commemorate the hour of his death? Some scholars have claimed that Robert Devereux was actually Queen Elizabeth's son by Robert Dudley, her intimate and dear friend since her youth. Did Dee know the truth of the matter and leave this entry as a clue? Devereux's trial had been on the 19th previous and Francis Bacon himself (also claimed by some to be Elizabeth's first son by Robert Dudley) had been very close to the case, trying to do whatever was in his power to save Devereux from the block. When the 1st Earl Walter died when Robert Devereux was a boy, Robert was given into the care of Robert Dudley, Queen Elizabeth's alleged 'eyes' and favourite, possibly the boy's actual father. Regardless, Devereux had always been near and dear to the Queen and was clearly another favourite. His beheading on 25 February caused great distress to those near him. The Queen is said never to have recovered. She could not have prevented it, it was out of her hands.

It is not my purpose to investigate the case of Devereux's alleged crime, but there is a curious detail surrounding the last days of Devereux's life. Inscribed in the wall of the stairwell entrance to the Beauchamp tower where Devereux was held as prisoner before his execution is an engraving of the name 'Robert Tidir.' Tower officials say that no prisoner of this name is recorded or known to history or tradition. So, the Earl may have made the engraving to tell the world that he was of Tudor Lineage. In the first edition of "The History of King Henry the Seventh" by Francis Bacon (1622) the name 'Tidder' is introduced in place of 'Tudor,' the House of Tudor being an English Royal House of Welsh origin. The Royal House commenced with the reign of Henry VII, the father of Henry VIII, and lasted until the death of Elizabeth I. If what these scholars claim is true, that Robert Devereux Earl of Essex, was the second son of Queen Elizabeth by Robert Dudley and Francis Bacon was the first, they would have been brothers and of Royal descent, potentially in the line of succession. If such were the case, Bacon's failure to save his brother might have created resentment by some who were close to the Earl.

Theodorus Dee died six weeks after the beheading of Robert Devereux on 25 February, on or about 9 April 1601. Already the date '9 April' had attracted my attention as the reader knows. It is the same day of the year in 1626 that the death date of Francis Bacon was registered to have occurred. The same day of the year is associated with the 3rd Earl of Pembroke's death. Pembroke was connected through Mary Fitton to Dee's visitor Mr. Holcroft. Mary was the young woman Pembroke refused to marry when she was expecting his child. I quickly realised that there could be 'hidden circumstances,' perhaps 'axes to grind' regarding Dee's final diary entry on 9 April 1601. I found it poignant and disturbing.

What actually happened to Tidder Dee on or around 9 April 1601? Dee's diary makes no mention of illness or accident, but by then advancing age was taking its toll. His entries were sparse. It is my opinion that something untoward happened to the boy. If so, there must have been a motive. I had spent some time over the years studying and collecting the works of Edward D. Johnson, a cypher man who had studied and published the works of Francis Bacon. Fortunately, I had a copy of his book, "The Bilateral Cypher of Francis Bacon" in my possession. Johnson's book was published in 1947. He had used the coded translations by Elizabeth Gallup (d.1934) the American scholar as one source for his book. I include a section from the last page:[3]

> From "The New Atlantis," 1635, Francis Bacon.
> I am named in the world, not what my style should be according to birth, nor what it rightfully should be according to our law, which gives to the first born of the Royal House (if this first born be a son of a ruling prince and born in true and right wedlock) the title of the Prince of Wales. My name is Tidder* yet men speak of me as Bacon, even those that know of my royal mother and her lawful marriage with the Earl of Leicester [Robert Dudley], a suitable time prior to my birth. Those whose chief desire is Scientia will rejoice in my experiments in Natural sciences, for they have greatly increased the knowledge which was in the world. Something have my labours done for other claimants and Philosophy and the Arts have gained by no means slightly by my labour, for I took no respite for years.

Mr. Holcroft's visit to Dr. Dee (cited above) was for a purpose unrecorded and the entire story seemed untold thus far, but clearly Theodorus 'Tidder' was a significant part of it. Although John Dee claimed to be descended from Roderick the Great, Prince of all Wales, and King Arthur, he never used the name 'Tidder' in connection with any (other) of his eight children. So why was Theodore associated with the House of Tudor in his Manchester burial entry? I returned to the entry of the birth of this child on 28 February 1588 in Dee's diary (Fenton edition). The entry reads as follows:

> Natus Theodorus, etc. In the morning, a little before sunrise. Theodorus Trebonianus Dee was born, with ascending in his horoscope. Lord's Day.

It seems that the birth had taken place in Trebon forty weeks after a contract between John Dee, Edward Kelly (his companion on the journey) and their two wives had been agreed to. The agreement was that the couples, upon spiritual advice and divine intervention, were to engage in 'marital cross-matching.' Edward Kelly's wife, Joan Cooper, seemed to be incapable of becoming pregnant and Kelly was anxious to become a father. Dee's wife Jane Fromand appears to have clinically fulfilled that role. Theodorus was regarded by Dee as 'a gift of God.' It was therefore surmised and in Court circles understood that Dee's wife Jane was the mother of Edward Kelly's son. The written and signed contract was expected by them to demonstrate that it was a responsible act between two mature married couples and that they had agreed to it.

In Dee's diary the dates immediately before and after the birth indicate there was a flurry of recognisable names in the entries. For example, "Mr. Francis Garland and his brother Robert went from Trebon" on 4 February. In the same diary entry Dee reports that he wrote to 'Mr. Dyer.' It will be remembered that Dyer was an agent for Queen Elizabeth and godfather to Dee's eldest son Arthur. I concluded that 'Garland' was a pseudonym for William Stanley, the 6th Earl of Derby who was responsible for Dee being appointed as Warden for the Collegiate Church in Manchester. Both Garland and Dyer had been frequent visitors to the Dee entourage whilst they were abroad.

The arrival of Theodorus Trebonianus Dee became an issue of debate in Court and ecclesiastical circles. The agreement badly affected Dee's reputation. On his return to Mortlake in late 1589 Dee was to find it increasingly difficult to establish himself in a paid professional role. Various positions were under discussion and after a considerable number of months and then years of delay and disappointment it was to the disagreeable position of 'Warden' in Manchester's Collegiate Church that he settled. He disliked the role, regardless of the power he gained within the church body. It was mutual.

And what of Edward Kelly? Having developed a successful career abroad, some years later he is said to have broken his leg in a fall in November, 1597 and died from the wounds, never to return to the British Isles. He was allegedly attempting to escape from confinement in prison, in protective custody. His earlier popularity had waned. I found the mysterious circumstances surrounding Theodorus increasingly perplexing and those of Edward Kelly just as much so. Dee's diary for March and April 1587 also includes the extraordinary entries regarding the 'new and strange doctrine,' which resulted in the cross-marital agreement and the birth of Tidder Dee.

Dee's first son Arthur was seven years old at the time of this 'compact' and his father was anxious for him to be trained as a scryer. Edward Kelly was to be part of that education process. The procedure included consultations with spiritual forces for guidance, help and support. Arthur was not responding to the training in a very positive way and it seems that this coincided with Edward Kelly learning that his wife was barren. On 18 April 1587 during yet another serious session of contemplation, which might also be called a séance, a spiritual entity by the name of 'Madimi' appeared before Edward Kelly. John Dee and his son Arthur were present. In Dee's diary the description is as follows:

> E.K: 'There appeared Madimi, II and the rest: and so they are here'. But now all the rest of them are gone, and only Madimi remaineth. Madimi openeth all her apparel and showeth herself all naked; and showeth her shame also.

It was during this session that it was expressly given to Kelly "That we two had our two wives in such sort as we might use them in common." Although protestations were expressed all round, it was made clear to Kelly by Madimi that it was the will of God. Quote: [4]

> E.K. "It appeareth written upon a white crucifix, as followeth:
> 'If I told a man to go and strangle his brother, and he did not do it he would be the son of sin and death. For all things are possible and permitted to the godly. Nor are sexual organs more hateful to them than the faces of every mortal. Thus it will be: the illegitimate will be joined with the true son. And *the East will be united with the West*, and the South with the North'. Now it is vanished away.
> Actionis Tertiae finis." (End of third action) [See pages 253-255]

Apparently, Dee provided a note in the margin of this highly significant entry:

> Arthur was smitten in a swound (swoon) and E.K. saw one in a long white garment make as though he would smite him. He was very sick for the time.

Seven years old Arthur had been profoundly affected by what he had heard and seen on the occasion. The covenant (agreement) was drawn up and signed on the 20 May by the two couples. On 21 May, Dee's entry reads:

Pactum Factum

Translated, it means 'agreement put into practice.' By 17 June John Dee noted that his wife Jane was showing signs of being pregnant. Theodorus (Tidder) was born the following 28 February 1588. Tidder was brought up with the rest of the Dee children, enjoying what would seem to be a normal childhood. After all, he was his mother's son and the family returned to England the following year without Edward Kelly. It seems that a few weeks after the birth the relationship between Dee and Kelly was beginning to deteriorate. An intriguing entry in the diary encouraged me to continue to have doubts about the situation in the Dee household. For 10 May (1588) Dee writes:

> Mr. E.K. did open the great secret to me, God be thanked.

Dr. Dee's family had their last meeting with Kelly on 16 February 1589, almost one year after Tidder's birth, when the family left for Prague. There were no further meetings between them whilst they were abroad. During that final year the diaries demonstrate that Mr. Dyer and the Garland brothers were again regular visitors to the Dee household. Queen Elizabeth was making sure that she was being kept informed of Dee's routines and programming.

Upon leaving Europe for the long journey to England with his family, Dee had hoped to meet Kelly at Stade. Dee had expected him to be returning to England about the same time on his own. This was not to be. But he did meet "Mr. Dyer coming to Stade around 17 November." Dyer was the last 'named' person in Dee's diary before taking ship, the Vinyard, on 19 November. They arrived at Gravesend on the English coast on 22 November 1589. Upon returning to his home at Mortlake, Dee found that his library had been ravaged by his supposed friends in his absence. Scientific equipment was damaged and spoiled. He complained to Queen Elizabeth and expected that in return for his loyalty to the Crown he would be given a dignified professional academic position.

It took until his Manchester appointment in 1595 for any real status to be given him. During the years intervening he received 'hand-outs' from the Queen. For instance, for Christmas 1590, having had yet another meeting with her on 27 November at Richmond Palace, her Majesty announced that "she would send me something to keep Christmas with." But this fragmented generosity was not good enough for the doctor and certainly not for his spirited wife Jane. They had a young family to maintain. As the years crept by the frustration of not having a regular income took a real hold. Dee was having to sell his books and precious artefacts to provide for his family. By that time I believe Dee himself had informed the Queen that Edward Kelly was in fact Francis Bacon. On 3 May 1595 Dee wrote in his diary;

> Between 6 and 7 after noon the Queen sent for me to her in the priory garden at Greenwich, when I delivered in writing the heavenly admonition; and her Majesty took it thankfully. Only the Lady Warwick and Sir Robert Cecil his Lady were in the garden with her Majesty.

Jane, the much younger wife of the learned doctor, on 7 December 1594 previous, had decided to take matters into her own hands. The frustration had been gathering some sort of momentum during the previous six months. Four years had passed since their return to England. Dee recorded this sobering entry in his diary:

> Jane my wife delivered her supplication to the Q.M. as she passed out of the priory garden at Somerset House to go to dinner at the Savoy to Sir Thomas Henedge. The L. Admiral took it of The Queen etc. Her Majesty took the bill again and kept her cushion.

The 'bill' in question must have been a copy of the 'pact' that had been formed, written out, and signed by Dee and his wife and Kelly and his wife prior to their cross-matching. Dee's infamous contract between the two couples regarding 'marital cross-matching' was now in the hands of the Queen. It must have been a shock. Over the next few days, the Queen having "referred all" to the Archbishop of Canterbury, Dee was required to have a meeting with the Archbishop before the Queen joined them at the Bishop's house. That was on 29 May. Four days later John Dee, his wife Jane and the seven children were being entertained by the Queen at Thistellworth. Jane kissed the Queen's hand.

A few days later Dee requested that the Archbishop of Canterbury should visit the family at Mortlake, which he did, and had supper with Dee at his residence. This was a few months later, after the 'cushion' incident on 7 December. Hereafter the drama began to accelerate. Four weeks after that event, on 3 January, the Wardenship of Manchester was first mentioned to Dee by the Archbishop of Canterbury. This was followed a few weeks afterwards by a meeting between Mr. Francis Garland (alias 3rd Earl of Derby) and Dee. In describing that meeting on 18 March, there was much talk between the two men of Edward Kelly.

It seemed to me that there had been some sort of intrigue under serious review during these months. The clues were becoming increasingly formidable in my considerations. By the evidence, I had come to suspect that Edward Kelly had confided to Dee that he was really Francis Bacon using the name Edward Kelly. Bacon had fooled Dee previously about his identity, Bacon being a good actor and master of disguise. Thus Theodorus, Kelly's son, conceived by a spiritually constituted agreement between the couples, was also Bacon's son. And since Francis Bacon considered himself to be the son of Queen Elizabeth, the 'virgin' Queen, Tidder, as a grandchild of the queen of England, could therefore be regarded as a future heir to the throne.

Francis Bacon and Edward Manners. Third Earl of Rutland.

VALE ROYAL
CHESHIRE
HOLCROFT FAMILY CONNECTIONS

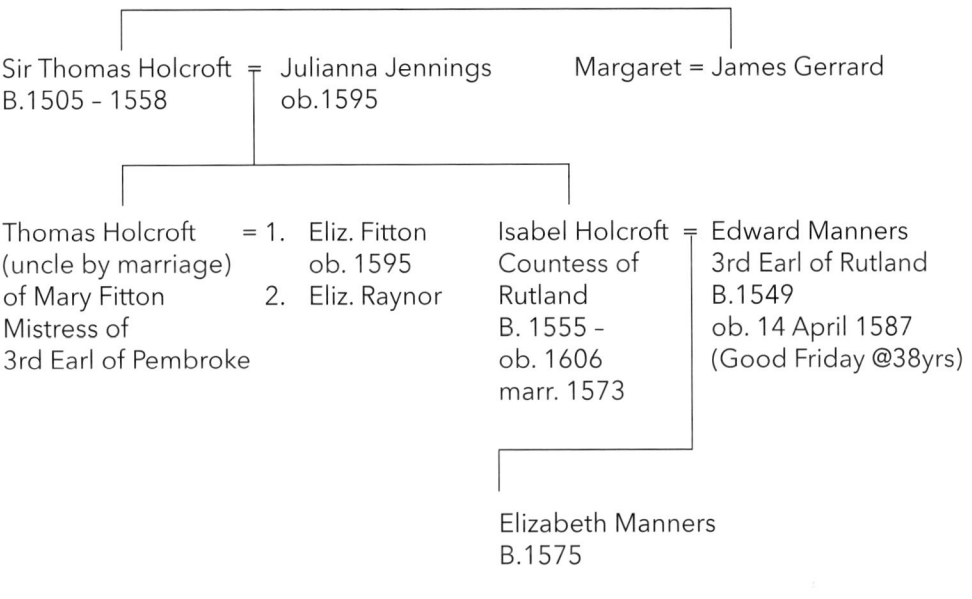

Sir Thomas Holcroft = Julianna Jennings   Margaret = James Gerrard
B.1505 - 1558   ob.1595

Thomas Holcroft   = 1. Eliz. Fitton   Isabel Holcroft = Edward Manners
(uncle by marriage)   ob. 1595   Countess of   3rd Earl of Rutland
of Mary Fitton   2. Eliz. Raynor   Rutland   B.1549
Mistress of   B. 1555 -   ob. 14 April 1587
3rd Earl of Pembroke   ob. 1606   (Good Friday @38yrs)
                                               marr. 1573

Elizabeth Manners
B.1575

Ref. 1   John Dee's diary - Fenton/Bailey Entries
          Dee's last - April 9 (Monday) 1601.

"Mr. Holcroft" of Vale Royal - visited Dee in Manchester.

Ref. 2   Isabel Holcroft
          See 'Notes'. Chapter 10, note 7. (J.H.)

Portraits:
Francis Bacon. (The Royal Society, London)
Edward Manners. (Wikimedia Commons)

The 'Duplicitous' – Geneology – 'The Holcrofts'

This might have been quite a shock to Dr. Dee as it unfolded, quite like that given to Elizabeth the Queen.

It was the entry two days after the Francis Garland meeting on 18 March which caused me the most concern in my thoughts about the matter. The meeting took place on 20 March. The recorded note is as follows:

> Mr. Marmion Haselwood, Mr. Dymmok, Mr. Hipwell cam to me to Mortlake.

These were distinguished gentlemen indeed. They were official holders of the hereditary office of 'King's Champion.' The role of the 'Champion' was to ride into Westminster Hall at the coronation banquet and challenge all who might question the King's or Queen's title. Descendants of the Dymoke and Marmion families are still based at the Manor of Scrivelsby in the parish of Horncastle in Lincolnshire. They have held the position of Champion going back to similar positions in Norman times.

The status of these visitors to the Dee household would imply that there was a need to visit Dee in connection with a Royal 'issue.' This had to be in connection with hereditary rights. Why else would they need to visit Mortlake? Given the interest by the Queen in the Dee family and the visit by Dee's children to visit her, I believe that she was particularly interested in Theodorus, the child born of Edward Kelly and Jane Dee.

I had long been interested in the character of Edward Kelly. His 'confidential disclosure' to John Dee shortly after the birth of Theodorus was certainly worthy of the 'godly' comment from Dee if the following is true: Did he admit to Dee that he was really Francis Bacon? This was on 10 May 1588, a few weeks after the birth of Theodorus. Given that Francis Bacon claims himself that he was the eldest son of Queen Elizabeth, if Theodorus was his son, 'Tidder' was also the grandson of the Queen. If this were the situation, no wonder the meeting at Mortlake took place.

But how could Edward Kelly, alias Francis Bacon, carry out such a charade for so long? I had been suspicious of how it could be done for some time. Clearly, he must have had a double, a not uncommon practice throughout history. This 'double' carried out Bacon's official duties as MP etc. for as long as was necessary when the need arose. But who had sufficient gravitas to carry out such a role for Bacon? I believe it was the 3rd Earl of Rutland, Edward Manners. Images in portraits of the day show them to be almost identical in demeanour and poise and very similar in appearance.

Moreover, as my investigation progressed, I began to suspect more and more and have since come to believe that Francis Bacon was the author of Shakespeare's Sonnets, as well as the 'main-spring' of his plays. Significantly, the Earls of Rutland, the 'Manners' family, owned Fisher's

Folly at the time that the 3rd Earl of Pembroke met his untimely demise, 9 April 1630. Now I reach for my Shakespeare's Sonnets. If what I believe is true then surely there must be a clue within the mysterious collection of poems? I did not have to look very far. I quote in full 'Sonnet 39':

Sonnet 39

> O how thy worth with manners may I sing,
> When thou art all the better part of me?
> What can mine own praise to thine own self
> Bring? And what is't but mine own when I
> praise thee? Even for this let us divided live,
> And our dear love lose name of single one,
> That by this separation I may give
> That due to thee which thou deserv'st alone.
> O absence, what a torment wouldst thou prove,
> Were it not thy sour leisure gave sweet leave
> To entertain the time with thoughts of love,
> Which time and thoughts so sweetly doth deceive,
> And that thou teachest how to make one twain,
> By praising him here who doth hence remain!

The use of the word 'manners' in the first line seems unusual but telling. The 3rd Earl of Rutland was Edward Manners (1549 – 1587). He was married to Isabel Holcroft, the sister of Thomas Holcroft, the guest of John Dee on that memorable if infamous 9 April 1601, the date of Tidder's death and the end of Dee's diary. Maybe to some, this interpretation is an unlikely scenario, but for me this 'Sonnet 39' holds other resonances. The Sonnet speaks of 'divided live.' (Line 5). The last two lines in the piece suggest living separate lives while the first two lines speak the truth of the matter. The Sonnet strongly implies a duplicity. While it may be thought that this sonnet is a simple love poem addressed to 'another,' it may also be interpreted as a lament addressed to one's 'self' and one's own circumstances, the thrust of all Shakespeare's work. Shakespeare challenges us to see ourselves through our fellow's eyes through God's eyes. Not an easy task. Shakespeare was anything but 'simple.'

At the same time there is another clue. Anne Bacon, Francis's surrogate mother, was Lady-in-Waiting to Queen Elizabeth before Elizabeth's ascension to the throne and was later responsible for translating the thirty-nine articles from Latin of John Jewel's "Apologie of the Anglican Church" published in 1564. Written in the year after the Church's inception, it was the historic defining statement of Anglican doctrine, separating it from the teachings of Roman Catholic and Calvinist churches. A historic time in the history of religion.

Francis Bacon cared for Anne Bacon during her last years at his property Mount St. Albans in Monmouthshire in Wales. She died in 1610. I believe that she was buried at St Michael's Church, Tintern, for which the 3rd Earl of Pembroke, William Herbert was Patron at the time. Ann Bacon was carrying out a duplicitous role as 'mother' to Francis Bacon, who I believe was really a Tudor, the son of Elizabeth. 'Sonnet 39' carries its own message, as does Sonnet 72:

My name be buried where my body is.

I believe that Francis Bacon was also buried in St. Michael's in Tintern. It seems to me that Theodorus, 'Tidder Dee' met an untimely death in Manchester on 9 April 1601, for mysterious reasons. By whose hand? That now becomes another question. The historical background of the Earls of Rutland during the Elizabethan period is certainly colourful in itself. Concentrating on Edward Manners, the 3rd Earl of Rutland, (the double) he, having married Isabel Holcroft in 1573, had one child, a daughter Elizabeth who inherited the Barony of de Ros. When the 3rd Earl died in 1587 he was succeeded by his brother John Manners as 4th Earl. This 4th Earl survived his older brother only one year, dying on 24 February 1588. It is said that because of his older brother's other commitments it was John who managed the family estates. At the time, the 3rd Earl Edward's official commitments were many and varied, while Francis Bacon's official duties were few and sporadic.

As noted, the Dee party left for the Continent on 23 September 1583 with Edward Kelly and his wife as part of the company. The 3rd Earl of Rutland was already living a very expensive and lavish lifestyle by that time. Early in life, the Earl had been made a ward of Queen Elizabeth, his father having died in 1563 when he was fourteen years old. On Shakespeare's birthday, a few months after the party had left England on 23 April 1584, this Earl was made a Knight of the Garter and a little later in 1585 Lord-Lieutenant of Lincolnshire. The Queen promised that he would succeed Sir Thomas Bromley as Lord Chancellor. Bromley died 12 April 1587. Rutland succeeded him for two days, dying himself on 14 April. It was but a brief promotion.

The family seat of the Earls of Rutland is Belvoir Castle in Leicestershire. Parts of it remain occupied by the family to this day. Despite what seems to have been a complex lifestyle during this time in their history, the 3rd and 4th Earls, who were brothers, died within months of each other, the elder acting, in my opinion, as 'pretender' (double) for Francis Bacon, which contributed to his absence from his estates. Both brothers were dead before the birth of Theodorus (Tidder) Dee.

The 5th Earl of Rutland was Roger Manners. He was the son of John, the 4th Earl. Significantly he was married to Elizabeth, the daughter of Sir Philip Sydney. Some historians have suggested this Manners was the actual author of Shakespeare's plays. Dying in 1612, it was his brother Francis who succeeded him. Francis was the owner of Fisher's Folly in 1630, hosting no doubt the crucial celebrity dinner for the 3rd Earl of Pembroke. But Francis, 6th Earl of Rutland had taken part in the rebellion in 1601 with Robert Devereux, the 2nd Earl of Essex, which had resulted in Essex's execution. Francis Manners was sent to prison and fined. Memories linger. Did Francis Manners have a role to play in Holcroft's visit to Dee regarding the 'happenings' in Manchester on 9 April 1601, at the time of Tidder's death? Coincidental or not, the demise of the 3rd Earl of Pembroke, William Herbert, is recorded as occurring on that same day of the year at Fisher's Folly in 1630, after the celebrity dinner on Rutland's property. Herbert was included by name in the Dee's diary 9 April 1601 entry. All these people knew each other well, were associated in Manchester and Tintern in Wales, and had their own parts to play in this magnificent story of intrigue.

Of course, Francis Bacon had his own part to play, including a significant role in the Essex trials of 1601. The tragic outcome of the trials and the death penalty for Essex, Bacon's brother, may have resulted in Theodore's own loss of a future life, potentially as heir to the throne. The question remains. The date 9 April carried implications for many years, stepping-stones through history, as we have seen. But there was one more stepping-stone to tread, yet unnoticed in this journey of enquiry. John Dee remained in Manchester until the death of his wife Jane on 25 March 1605. It is thought she died of the plague. By then Dee's younger children had died as well of the same disease, which was rampant in the town. His eldest daughter Kathryn survived and helped her father move back to Mortlake. Arthur was by now married and no longer at home. Significantly, he was sent away to Russia on a 'diplomatic' mission, perhaps to remove him from the scene. Did he 'know too much?' Dee's popularity had not improved. He was thought to have practiced 'black arts' and even requested King James 1st to put him on trial for witchcraft. It never happened.

Dr. Dee died on 26 February 1609. His last years were shrouded in mystery. He was a man who had earned the highest respect for his expertise in mathematical and scientific scholarship. His medical and philosophical practices were sought-after all over Europe and his guidance in voyages of sea travel was much in demand. He is thought to have been buried in the Chancel of the parish church of Mortlake.

1609, the year of Dee's death, was the same year Shakespeare's Sonnets were published. As already noted, the author of the collected poems was, in my opinion, Francis Bacon. The dedication was to the 3rd Earl of Pembroke. Dee's death would have had a profound effect on both men. Dee had played such a significant role in their lives, as the Sonnets quoted here have suggested. Pembroke and Bacon in turn would have had a significant role in the decision-making of this remarkable man – John Dee. But where is the evidence for that? It was Dee's years in Manchester that provided me with answers to questions which I had been uneasy about before. I looked again at the diary entries prior to his move back to Mortlake in 1596. Dee had arrived in Manchester on 15 February that year. But his wife and children left Mortlake for Coventry on 26 November 1595. The day before that event Dee records "News that Sir Edward Kelly was slain," which information turned out to be incorrect as recorded by Fenton. There followed a three months gap in the family's communal domestic arrangements. It is my belief that Jane and the children spent at least some of the time at the residence of William Overton, the Bishop of Lichfield and Coventry. Overton had been appointed Bishop in 1580 and shared a unique interest with Dee. The Bishop, apart from his clerical duties, was responsible for developing glass making in Staffordshire. Dee had visited glass-making premises during his years abroad. He needed vessels for his experiments. A diary entry for 11 May 1586 is worthy of inclusion at this point:

> I came to Leipzig and was at Peter Hans Swartz his house lodged. I found Lawrence Overton (with much ado), an English merchant: to whom my wife (the last year) had shown no little friendship to himself, and Thomas, his partner's servant, in the time of his lying sick in our house, etc., at Prague.

A few days earlier (6 May) Dee recorded that he had visited the Valkenaw 'glass-house.' The surrounding entries in the diary indicate that the Lawrence Overton mentioned in the entry was part of the glass-making industry and that he was a kinsman of Bishop William Overton, who himself had a singular interest in the development of glass making. In any event, it seems that Jane Dee and the children had spent time in what may be described as a safe ecclesiastical environment before settling into their new Manchester surroundings when Dee became Warden of the church.

Many years after this line of inquiry, a consultant of mine, Peter Welsford, drew my attention to a rare book which he thought might be of interest to me. I bought it and read it. Deciding that it was not relevant to my studies at the time I put it to one side. Eventually, having found myself curious about this Coventry 'haven,' and after more recent findings, I looked at the book again and recognised what I can only describe as the final stepping-stone in this part of my journey. I think it appropriate to include Peter Welsford's own account of how he came across the rare book he suggested might be of interest to me:[5]

Apart from the usual duties of Treasurer and Secretary of the Scientific and Medical network, a UK based international organisation of Scientists, Doctors, Philosophers and other professional academics whose members look a little beyond the norm ("outside the box")... and in the run up to The Millennium (2,000 AD), I had also become interested in The Francis Bacon Society (founded by Mrs Pott, and said to be the oldest Literary Society in this country) quite by chance – having come across a little booklet titled: "How to Crack The Secret of Westminster Abbey, (Time Brings Forth The Hidden Truth), by Richard Barker, first published, 1986.

All credit to Thomas Bokenham ("Bokie") an ex-Bank of England official who had soldiered on as Chairman for many years virtually single-handed and, with Richard Barker, had succeeded in cracking the Secret, found within the near life size statue of William Shakespeare to be seen in Poets Corner in The Abbey, where WS is leaning with one hand on a pile of books whilst on the other, his finger is pointing to a Scroll – bearing a quote from The Play, The Tempest – starting with the lines…"Shall Dissolve, and like the faded fabric of a vision … " and to the word: Temples.

There follows a detailed explanation of the decoded words, namely "Francis Bacon", which emerges arched, in the centre of the Scroll perfectly symmetrically.

Next the BBC became interested in this discovery thanks entirely to Gwyn Richards a Senior Producer at Pebble Mill, who organised a meeting between Bokie, Richard Barker, the Abbey Librarian and me, as the Presenter of an unscripted meeting in Poets Corner in The Abbey; which later went out 'live' several times on BBC World Service.

It was here, in the Abbey, I had met Bokie for the first time and after attending a few of their regular Talks and Meetings I was invited to join the Council and later, became Editor of Baconiana, the in-house Journal for several years. Bokie had also decoded many of The Sonnets which he published in a little booklet. Towards his later years, and together with Gerald Salway the Treasurer and Secretary, we used to take him out regularly for lunch at the Bank of England Sports Club in Putney, until he passed away in 2003 when I was appointed Chairman of the Society. When Gerald and I were sorting out a mass of books and papers with the help of Rubie his widow, we ended up in his bedroom – with a portrait of Bacon above the bed head but beneath the bed, there was a copy of a booklet titled:

'The Burial of Francis Bacon and his Mother in The Lichfield Chapter House … concerning Rosicrucians'.

Written by Walter Arensburg and published in 1924, Arensburg believed he had discovered the whereabouts of Francis Bacon's burial place, after an elaborate series of researches spelt out and carefully detailed in the book, even confirmed by a reference to Cymbeline and Jupiter's Label, the bookplate to The First Folio.

Armed with this information he came to UK and visited The Cathedral, but his request to be allowed to investigate the site in The Chapter House was rejected by the authorities – hence, it seems, they prefer their continuation of – The Status Quo.

I have included Peter Welsford's account in full to demonstrate how research findings can emerge as a result of due diligence, loyalty and trust.

Bishop William Overton died on 9 April 1609. 9 April, again, that fateful date. An eerie reminder. John Dee had died a few weeks before on 26 February. I think it may well be that the two burials occurred in the eight-sided Chapter House of Lichfield Cathedral, so carefully researched and recorded by Walter Arensburg. I believe it was those of Theodorus (Tidder) Dee and his mother Jane Dee, having been transferred from the Manchester Collegiate Church to the diocesan Cathedral by John Dee after Jane's death in 1604/5 (exact date unknown). That would have been before he finally left Manchester for Mortlake. Were mother and son buried in a replica of the contemporary 1599 eight-sided Globe designed by John Dee for London? (As interpreted from the Byrom collection of geometric drawings).

It may be said by some that my conclusions are a figment of my imagination. That is understood. But my suggestions are based on the citation of accumulated facts from rarefied information not available until recently.

As a final note for this part of the journey, I include the following quote from "The History of the City and Cathedral of Lichfield" by John Jackson first published in 1805:[6]

In the year 1604 the Earl of Nottingham, the Earl of Suffolk, and others wrote to Ashmore Esq. high bailiff, and the corporation of Lichfield, in order to procure from the Bishop (William Overton) the fee farm of the manor and lordship of Lichfield, for the earl of Essex.

This Earl of Essex was the son of the unfortunate Robert Devereux the 2nd Earl, who was executed in 1601 for treason. Also named Robert, he was thirteen years old when King James 1st chose to elect him 3rd Earl in 1604.

He would have been the same age as Theodore (Tidder) Dee in 1601 who died on 9 April and he would have been the cousin of the 3rd Earl of Essex if his father and Francis Bacon were brothers. The gift of the fee farm of the manor and Lordship for the Earl of Essex could be seen as a gesture of respect towards the burials of mother and son in the Chapter House of the Cathedral and the execution of Elizabeth's favourite, Robert Devereux, 2nd Earl of Essex, who's surviving son, the 3rd Earl and Tidder would have been Queen Elizabeth I's grandchildren. Elizabeth died in 1603.[7]

CHAPTER ELEVEN

Telling moments.
The Tintern Allegory.

ALTHOUGH I HAD gathered a considerable amount of original historical data over the years, a large amount of it had already found its way into my published books. Some of it had not, fermenting in a storage area in one of my study rooms, awaiting a 'lightning moment' to emerge.

The Reverend Neville Barker Cryer died on 2 July 2013. As mentioned, we had engaged in regular meetings over a twenty-three year period discussing unique elements of my research, often following Freemasonry and Templar principles. His opinions on some of it can be seen quoted here and in my earlier books. He was a distinguished and much respected Masonic author in his own right, as well as a Reverend of the Church of England. I had been introduced to him by Richard Leigh, one of the authors of "The Holy Blood and The Holy Grail."

After some time, my collection of the Reverend's works became considerable. Files of recorded meetings, letters, papers and books persuaded me after his death to form an archive for easy reference in one study room. It was on one occasion of sorting out his correspondence that I came across a letter with contents I had put to one side. I felt at the time they were not relevant to work in progress. But at this juncture I was gripped by the contents – three pages of a 'paper-back' book which included part of an illustration from a painting. There were no explanatory details with the enclosure. I tracked the pages to a novel entitled "Landscape of Lies"[1] by the distinguished art expert and author Peter Watson. The title was an echo from the past.

The complete painting was displayed in a 'fold-out' at the front of the book. The novel is a detective story, an investigation carried out to learn the provenance of the painting and to interpret its meaning as a 'map' to discover 'treasure.' The book was first published in 1989. My attempts to find the actual provenance of the painting resulted in my being told that the author had used it to 'spin his yarn' very successfully. It was sold. The details of where the sale took place and to whom were not known.

I was immediately captivated by the features in it. The excerpt sent to me earlier by Rev. Cryer was but a section of a much larger picture. I was mesmerised. I read the novel from cover to cover and made notes of the interpretation by the author and his understanding of certain features in the painting. The story was intelligently written and there was much to learn about the 'art world' from the author, who clearly knew what he was talking about. The picture is of a landscape, clearly allegorical in nature, containing human figures and objects and structures I recognised. I believe this 'allegory' is set in the landscape of Tintern, which differed from the site described in the published novel. But, there is an Author's note printed at the front of the book which reads:

> This is a work of fiction. All of the characters
> And some of the locations are inventions.

As time went by, I became convinced that my recognition of the setting of Tintern was correct. Consequent to that recognition, the picture told me a story different to the one in the novel. The painting tells 'truths' that I have come to consider of great importance, even if I do not fully understand all of them from my studies over the years, some which are recorded in earlier chapters of this book.

The figure of a man lying on the ground with a tree growing out of his navel represents the 'Stem of Jesse', meaning of genealogical descent from the biblical House of David. I believe this figure depicts Francis Bacon. I am also convinced that the window feature in it is part of the architecture of Tintern Abbey. It is similar, if not identical to the Tintern Abbey window depicted on the cover of my book "The Hidden Chapter." Although I had not seen this allegorical painting before 2017, the artist and I used the same feature for different purposes at different times.

I have stated that I believe Francis Bacon was buried at St. Michael's Church in Tintern as the reader knows. It is also my belief that the altar depicted in this painting is that of St. Michael's church, placed allegorically within the environs of the painting and Tintern Abbey. Once that idea became fixed in my mind the rest of the painting presented astonishing facts, giving credence to my research and my interpretations of history. The cavalcade of figures, represented visually almost like sentinels, give up their silent rhetoric one by one. Their story is that of a historic drama. The artist clearly knew secrets that he/she was determined to display on record in a single piece. Who had acquired such information to be transmitted this way?

The Tintern Allegory – Shakespeare connections

The Tintern Allegory – The Geometry

My belief is that it was Sir Thomas Herbert of Tintern, an accomplished artist himself, the travelling companion of Sir Thomas Dodmore Cotton on his sea-faring adventures to Persia, Africa and Asia in 1626. This Thomas Herbert was later the permanent cell-mate for the last months of the life of King Charles 1st, brother of Elizabeth of the Palatinate, and son of King James 1st. Thomas Herbert was trusted by these-high powered individuals and would have been given confidences that he held in memory awaiting a suitable venue for sharing. He had acquired by then the Hermetic College in Tintern in 1640, now called The Nurtons, which had been a secret haven for scholars of hermetic studies for many years. The property was already in the possession of others in the family before him. As I say, Sir Thomas was a talented artist, as the illustrations in his published work of 1677 demonstrate, "Some Years Travels into Diverse Parts" of his naval adventures with Dodmore Cotton.

However, I do not think that Sir Thomas Herbert was responsible for the painting as it is today. There must have been a draught form of his in preparation already on the site. I think the 'finished' painting was completed by Oliver Lodge junior, the eldest son of Sir Oliver Lodge, the physicist, who was also an artist. This junior Lodge was in residence at The Nurtons from 1912-1918. Sir Oliver Lodge senior had rented it from the Rev. Clement Crutwell; Crutwell had purchased The Nurtons in 1877. Lodge Sr. had rented the property for his son and wife, Winifred Atkinson known as Wynlayne. Lodge Jr. wrote the book "What Art Is" while living in Painswick to commemorate his 50th birthday, dedicating it to his wife who had died in 1922. They had moved from Tintern to Painswick in 1918. 'Lost' historic data, I believe, has been stored at The Nurtons for centuries.

What then, is my interpretation of the story behind the painting of this carefully articulated piece of visual drama? The characters spread across the canvas are individuals who carry and/or wear a singular object which is intended to help identify them. Who is the 'torch-bearer' of it from my perspective? It is the bearded character on the extreme left, dressed in red. He holds the long rod which pierces or severs the arm of the merman figure in the pond.

This figure became the focus of my thoughts. I identify the victim as Merovee, the mystical figure who was by tradition the originator of the Merovingian line of Frankish Kings reigning from the early 400's to the mid 700's. A supernatural half-human and half-marine creature, he possessed superhuman powers and was responsible for much of the mystique surrounding the Merovingian dynasty. His line of rulers was all powerful in Europe until the mid-eighth century. My early sources had commented on Tewdric, a member of the Merovingian dynasty who allegedly lived in a cave underneath or close to St. Michael's Church until he was mortally injured. He was buried in Mathern Church, near Chepstow.

I need to remind the reader here of a biblical quote from St. Matthew, Chapter 5, verse 30, taken from the 'Sermon on the Mount':

> If thy right hand offend thee, cut it off, and cast it from thee.

It appears that the robed figure in red is indicating that Merovee, via his Merovingian descendants, had in some way violated the convention of probity. Could it be the pact between John Dee and Edward Kelly, aka Francis Bacon? Given that the gospel of St. Matthew describes Jesus as being of royal blood, he was a descendant of King Solomon and David, who emerge later, associated with the folklore of the Merovingians. He must then be regarded as a priest King in his own right.

The British Royal House of Tudor also regarded themselves as descendants of David, thereby connecting themselves to the Merovingian lineage by blood. Since Francis Bacon considered himself to be a member of the House of Tudor, he would then have been 'of the blood' if his mother was in truth Queen Elizabeth. In other words, Francis Bacon was a descendant of the Tree (stem) of Jesse, Jesse being the father of David. All this can be drawn out of a single picture painted on canvas by those who understand the nature of an allegory, the master of which is known to be Jesus, a descendant of Jesse. Alas, it is Bacon.

But who is the dominant red-robed figure boldly drawing attention to a flawed heritage and for what reason? His red gown provides an important clue. Facing him, but behind Merovee, is a woman displaying within the skirt of her formal dress a half concealed matching red stocking. The same colour red is used nowhere else in the painting apart from behind the head and shoulders of the prostrate figure of Francis Bacon (Jesus, the stem of Jesse) leaning against the pillar. Behind his head is a small red cushion giving him rest. The colour red binds the three characters together in a symbolic way. I remind the reader that in an allegorical painting the use of colour-coding is a device used by the artist to link elements of a story into an intelligible narrative. In his hand, Jesus (Francis Bacon) bears the holy-grail. The blood within the grail is a sign, the hidden red. The life and light, and bread – the flesh of God, transformed.

Studying the setting surrounding these figures I arrived at the conclusion that this part of the picture is representative of Shakespeare's play 'The Tempest[2]' – the red robed figure 'Prospero' is using his rod (staff) in a judgmental way. It has been traditionally accepted by many scholars that Prospero is based on John Dee, the Elizabethan Magus. In that event the red-stockinged female would be Jane Dee, his wife. The Francis Bacon figure resting against the red cushion is linked to Dee and his wife Jane. If one follows the pillar that Bacon is lying against to the top, figuratively displayed are the figures of Adam and Eve. But alongside them is another figure of a youth, holding in front of him a smaller rod, pointing downwards. Who then is the youth, positioned at the top of the pillar, supported by the cushioned head of 'Francis Bacon.' Tidder Dee?

'The Tempest' was first performed on 'Hallomas night' as entered in the official Revels Account of 1611. The venue was "at Whitehall before the Kinges majestie." Hallomas is usually 31 October. Part of an ancient Celtic religion, it was designed to scare away the ghosts and spirits at the end of the summer season. The play was next performed as one of a group of plays in celebration of the wedding of Elizabeth, daughter of James 1st to Frederick, the Elector Palatine of Bohemia in central eastern Europe. Frederick and Elizabeth married in 1612.

'The Tempest' is a play about magic, love and forgiveness, set on a remote Island. 'Prospero,' a magician with enormous powers, had suffered badly, having been at one time the Duke of Milan, now deposed. Though tempted, he does not take revenge but ultimately offers forgiveness. The play also examines the consequences of emigration to America and Bermuda. Members of the Virginia Company, formed in 1606 were responsible for those pioneer voyages, some of them disastrous. Investors in the company included such aristocrats as the 3rd Earl of Pembroke, William Herbert.

This is not the place to give a complete resume of the storyline, but having studied the painting at some length and knowing the substance of the play, I need to quote two speeches which are relevant. The first is spoken by 'Prospero.' It is taken from Act 5, lines 48-57:

48 Graves at my command
49 have waked their sleepers, oped, and let 'em Forth
50 By my so potent art. But this rough magic
51 I here abjure; and when I have required
52 Some heavenly music – which even now I do
53 To work mine end upon their senses that
54 This airy charm is for, I'll break my staff,
55 Bury it certain fathoms in the earth,
56 And deeper than did over plummet sound
57 I'll drown my book.

The second quote is from the same Act, Scene 5, lines 88-94. It is usually a song, sung by in part 'Ariel' (an 'airy' spirit).

88 where the bee sucks, there suck I,
89 In a cowslip's bell I lie;
90 There I crouch when owls do cry;
91 On the bat's back I do fly
92 After Summer merrily.
93 Merrily, merrily shall I live now
94 Under the blossom that hangs on the bough.

The two quotes I believe are represented in the painting. Wild-flowers, growing in the foreground have amongst them 'cowslips' and 'oxlips.' Most significantly, there are books and papers carefully positioned round the barrel nearby ("I'll drown my book"). I have identified 'Ariel' as the bat-like creature positioned between the red clad figure and Merovee. The monk standing beside the barrel with his face concealed and hidden is standing on a group of tiles. The tiles have patterns of swans on them. Under a magnifying glass it may be seen that the feet of the monk are in fact cloven, as for pigs, cattle, goats and sheep. Again bacon.

Shakespeare was known in his own time by fellow authors as the 'swan of Avon.' Here, the hidden face and cloven hooves imply that the artist is using pig's split two toes as a clue. The Prospero quote "I'll break my staff" completes in part this section of the cameo. But it is a third quote from the play which I find equally compelling and must be included because of the painting's faceless monk. It is taken from Act 1, Scene 2, Lines 269-284. Again, Prospero is speaking:

269 This blue-eyed hag was hither brought with child
270 And here was left by th' sailors. Thou, my slave,
271 As thou report'st thyself was then her servant
272 And for thou wast a spirit too delicate
273 To act her earthly and abhorred commands,
274 Refusing her grand hests, she did confine thee,
275 By help of her more potent ministers
276 And in her most unmitigable rage,
277 Into a cloven pine, within which rift
278 Imprisoned thou didst painfully remain
279 A dozen years; within which space she died
280 And left thee there, where thou didst vent thy groans
281 As fast as mill-wheels strike. Then was this island –
282 Save for the son that she did litter here
283 A freckled whelp, hag born – not honoured with
284 A human shape.

I have noted that in the painting 'Ariel' is close to the 'Prospero' figure (Dee). Authors have decided in their editing of the play that Ariel had been trapped by the character of the witch 'Sycorax' in a 'cloven pine' as punishment for resisting her commands. The small image of a witch-like figure can be seen in between the two end rings of the curtain rail above the altar on the right-hand side of the painting. The received explanation by scholars of the 'Ariel' (character) links the witch, and then the faceless monk (Shakespeare/Bacon) in the painting.

Bacon, whom I consider to be the author of the plays, was citing Sycorax as Queen Elizabeth I, his mother. The late Jean Overton Fuller – Bacon's biographer quotes from the States Papers Domestic :

> "The Spanish Ambassador – De Feria had confidential conversations with Lady Sidney, Leicester's sister, but in November, 1559, reported the Duke of Norfolk had threatened that if Lord Robert would not abandon his pretension to the Queen's hand he would not die in his bed.

> In mid-June a woman of Brentwood gossiped that Leicester had given the Queen a petticoat. Another exclaimed, "Thinkst thou it was a petticoat? No, no, he gave her a child, I warrant thee". A Mother Dowe repeated this in another village, expressing it that "Dudley and the Queen had played at legerdemain together and he was the father of her child." Another answered her, "Why, she hath no child yet." Mother Dowe reported, "No, and if they have not they have put one to the making." The report containing these phrases was sent by Lord Richard Thomas Mildmay to William Cecil (Lord Burghley). For Cecil the affair was alarming, and he always tried to keep Leicester back."

Francis Bacon was born on the 22 January 1561. John Dee had decided the coronation date of Queen Elizabeth 1st – 15 January 1559. I believe that he served as her "protector" in these early years and throughout her reign. Bacon, (Ariel, figuratively speaking in the play) had been trapped into this anonymous/silent position as an author because of his true biological parentage. Politics and expectations trapped him. There were a few contemporaries of Bacon and subsequent descendants who knew of his plight. Clearly the artist of the painting was one of them, as was 'Prospero.'

Prospero's use of his staff to cut the arm of Merovee is such a deliberate act in the painting that there has to be more information in support of that action somewhere. A Merovingian presence in historic terms is not unfamiliar in the Tintern area. The nearby Usk Priory was dedicated to the sixth century Merovingian Frankish queen 'Radegunda.' She founded the Abbey of the Holy Cross at Poitiers in France, having fled from her cruel husband Clotaire I. There, she became a nun. Frenchmen have travelled to Usk regularly at Easter and Whitsuntide to pay homage to her. Her charitable works were well known. She became one of the first Merovingian Saints, dying in the year 587, within eight years of Tewdric.

The Merovingian link becomes even stronger when the distinctive vase with an eagle on it resting on the St. Michael's altar on the extreme right of the painting became a focus for my attention. I understood it. It is a depiction of an ancient 2nd century Egyptian porphyry vase which has acquired the name 'Suger's Eagle.' Suger was the abbot of Saint Denis outside Paris, the resting place of many Merovingian Kings. Abbot Suger lived from 1081 – 1151 and is described as a French abbot, statesman and historian. A friend and confidant of French Kings, he is credited with rebuilding the "great church of Saint Denis, the burial church of the French monarchs."

Resident in the environs of the church from childhood, he had found "lying idly in a chest for many years an Egyptian porphyry vase admirably shaped and polished." He arranged for the vase to be set into a liturgical vessel "in the form of an eagle." Inscribed around the base is a dedication to the church of Saint Denis. The vase is now, along with other precious relics of the period, in the Galerie of Apollon at the Louvre in France.

The vase is unique. Abbot Suger would have been very cognisant of the Merovingian burials and the legacies connected with their lives and traditions. The artist must have seen the vase because the details are exact. However, there is one marked difference which must be considered deliberate. The actual Suger vase is coloured amber. In the painting it is blue. All the other details are accurate. I believe it is a deliberate echo of Bristol Blue Glass, famous for its striking colour and quality. I think the artist has used the colour out of respect for John Edmonds Stock who had been involved in the French Revolution in the late 1700's. He was accused of treason and escaped the guillotine by fleeing to Philadelphia in America. He completed his medical studies there, returned as a qualified doctor, set up a practice in Bristol and worked at the hospital of the Bristol Royal Infirmary. A descendant of Thomas Herbert, he was forced to forfeit his Tintern Parva Estate to the Duke of Beaufort, but in 1808 reclaimed it and was granted The Nurtons, with its acquired religious links to the nearby St. Michael's Church.

The Suger vase included in the painting comes as a 'knowing' clue of the heritage of The Nurtons and its Merovingian French links. It tells us that the finished painting is not in itself a Tudor painting but closer to the 20th century in its finished form. The Suger vase links St. Michael's Church with St. Denis, Paris. Already the church was linked to Tewdric.

Now we must turn to the central figure of the monk with his back turned from Bacon. He represents the Reverend Roger Williams (1603-1683), founder in 1636 of the Baptist Church in Rhode Island, America. He was a descendant of Roger Williams of the Lordship of Usk in Monmouthshire. A seemingly hostile figure, he stands alongside the hound of Hades, Cerberus in Greek mythology (the Odyssey), a multi-headed dog that guards the gates of the Underworld to prevent the dead from leaving. This feature demonstrates that the painting becomes sectionalised visually into two parts with underground features on each side. The composition reverberates the cloven or divided nature of humanity. "I'll drown my book," says Prospero. We can see manuscripts and books round the barrel on the left. The barrel is a feature referred to in another of Shakespeare's plays 'Richard III' (Act 1 Sc. 4). Quote:

> First Murderer. – Take that and that (he stabs Clarence) If this won't do the job, I'll drown you in the wine barrel in the next room. He exits with the body.

Richard III arranged to have his brother George, Duke of Clarence, murdered.

The results of the radar scans and other archaeological work commissioned by me with permission from the Diocese of Monmouth gave clear evidence of underground features at St. Michael's. For some years I have thought that St Michael's was the burial site for Francis Bacon and Ann Bacon (his 'adoptive' mother). See illustration Chapter Nine.

Roger Williams had close links with Bacon. His father James died when Bacon was seventeen years old. Sir Edward Coke, a barrister, became Roger's patron, supervising his education at Charterhouse and then Pembroke College Cambridge. In 1598 Coke married as his second wife Elizabeth Hatton when he discovered that Bacon was also pursuing the same lady's hand in marriage. Coke was also the main prosecuting counsel for the government in the case against Robert Devereux, Earl of Essex, Francis Bacon's brother, who, as mentioned earlier was found guilty and executed in 1601.

An earlier Roger Williams, part of the same Monmouth family, was a devoted, active supporter of Robert Devereux. He died in 1595 and Devereux attended the funeral before his own death. This Rev. Williams married Mary Barnard in 1629 and had come to regard the Church of England as 'corrupt and false." He and his wife Mary emigrated to America that same year. The historic Monmouth family seat of 'Williams' was Llangiddy. The earlier Roger Williams had bought the Priory of Usk in 1545 with its Merovingian links to Queen Radegunda. The family also held a reputation for its pack of hounds, which they carried on for generations. The artist, in positioning Rev. Williams with the hound of Hades, the three-headed dog reminds us of that family trait.

To consolidate the cameo supporting this identification I quote from "The Hundred of Usk" by Sir Joseph Bradney (Vol. 3 P.1):

Usk, Tregrog was part of the dower of two of the wives of King Henry VIII viz Anne Boleyn and Catherine Parr, the latter of whom obtained leave to sell it, And it was eventually purchased 1554-5, by Roger Williams of Usk.

I also include a quote from "The Church and Priory of St. Mary, Usk" by Robert Richards, published in 1904. It comprises documentation of official records, one of which is the following:

> In 36 Henry VIII, the site of the Priory and various lands particularly described were granted to Roger Williams, grandfather of Sir Trevor Williams. We are not told the price to be paid for these confiscated lands, but it would appear as if, either the royal robber was anxious to get rid as soon as possible of his ill-gotten acquisition or the said Roger had bribed the Auditer, Edward Gostwyk, appointed to value the lands, as he remarks in his certificate: 'Who wold purches the pmysses the Audytor knoweth not, but only this bringer!' The yearly value was assessed at £8, 4s 7d.

The Priory as we know was dedicated to St. Radegunda. In depicting Rev. Roger Williams in the painting, we now have three features with Merovingian ties: Merovee, Queen Radegunda and the vase identified as 'Suger's Eagle.' The position of these features within the structure is highly significant. The artist wished to highlight these facts. There must have been a motive for that not immediately evident in the design structure. Two figures, sitting dressed in brown, communicate significant information. The one to the left with the skeleton face and wearing a bishop's mitre is surely Thomas Cranmer, the Archbishop of Canterbury. He was burnt at the stake on 21 March 1556. I quote his last words:

> that unworthy hand...Lord Jesus, receive my spirit...I see the heavens open and Jesus standing at the right hand of God.

Cranmer had been consecrated Archbishop on 30 March 1533. Henry VIII had married Anne Boleyn on 24 February and Elizabeth, later Queen Elizabeth I was born on 1 September later that year. As part of her dowry Anne had been given the Lordship of Usk. Cranmer was very loyal to Henry VIII. His legacy lives on within the Church of England by the "Book of Common Prayer" and the 'Thirty-Nine Articles' translated by Anne Bacon, which explains those beliefs through a statement of faith.

Cranmer's inclusion, visually represented in the painting, is pivotal. A reminder of some of the events for which he was responsible will serve as confirmation of his power for many years:

1.a.i.   Cranmer baptised Elizabeth very soon after she was born and became a god parent.

1.a.ii.   Anne gave birth to a stillborn son in 1536. Henry became interested in Jane Seymour.

1.a.iii. Cranmer doubted Anne's guilt of sexual behaviour as accused. He saw Anne in the Tower of London and heard her confession. The next day Cranmer declared her marriage to King Henry VIII null and void.

1.a.iv.   Two days later Anne was executed (1536). Cranmer mourned her death.

1.a.v.   Henry VIII died on the 28th January 1547 after four further marriages.

1.a.vi.   Henry's teenage son was crowned King Edward VI aged nine years. His mother was Jane Seymour.

1.a.vii. In December 1551 the original Roman Canon Law of the church required revision. Cranmer formed a committee and recruited Peter Martyr, Jan Laski and John Hooper to participate.

On 6 July 1553 the teenage King Edward VI died and his sister Lady Jane Grey reigned for nine days. She was followed by Mary. Queen Mary I, daughter of Henry VIII by his first wife, Catherine of Aragon, reigned from 1553-1558. Mary was committed to returning the country to its old religion, Roman Catholicism. She was dismayed by the collective damage caused during the earlier 'change-over' to Protestantism to monasteries, churches, some of the dedicated clergy and others. Those who were the main participants in the Protestant 'rebellion' to the Catholic cause were charged with treason or heresy and burnt at the stake, as was the Archbishop of Canterbury himself, Thomas Cranmer.

The clergyman sitting opposite Cranmer in the painting I identify as Martin Bucer, the German Protestant Reformer. He was exiled to England in 1549 and under the supervision of Cranmer, influenced the second revision of the "Book of Common Prayer." He died in 1551. When Queen Mary I came to the throne she had Bucer tried posthumously for heresy. Found guilty, his burial casket was disinterred and his remains burnt along with his books.

The last figure, unidentified so far, is the dominant male standing in front of the green altar 'cloth' of the church of St. Michael's at the right of the painting. John Hooper, the highly controversial Anglican Bishop of Gloucester and Worcester was executed for heresy by burning on 9 February 1555. The artist wants us to be certain of that identification and provides several pointers. Born around 1500, Hooper as a young man was a member of the White Friars, a Carmelite friary situated on St. Michael's Hill in Bristol. The friary had been established in 1267 by the Prince of Wales who became Edward I. In the painting he wears a blue waistcoat with a St. Michael's decorative motif around his neck. The colour blue echoes the blue of the Suger Falcon vase, reminding us again of the famous Bristol Blue Glass industry. The White Friars Friary was surrendered to King Henry VIII in 1538.

Hooper was part of Cranmer's committee formed in 1551 to review Roman Canon Law, but he differed with Cranmer on major issues. He denounced the wearing of 'Aaronic Vestments' (church clergy apparel) and for a time refused to be consecrated according to its rites. Cranmer and Bucer and others urged him to submit. After a spell in Fleet Prison, he agreed to do so. But as a subscriber to the extreme Protestant view, Hooper was the first bishop to be attacked by Queen Mary. He was condemned for heresy, sent to Gloucester on his home ground and burnt.

It will be seen in the painting that though a bishop, Hooper is not wearing his 'canonicals,' and has sprouting from his ankles the wings of Mercury, reminiscent of the Greek God Hermes who wore winged sandals. He seems to be studying the Suger vase intently. The prominent eminences of Cranmer, Bucer and Hooper met untimely deaths as the result of religious conflicts during this time of extreme politics involving the Church and Crown. Each had close associations with the Tudor dynasty, which as can be seen in this painting had strong links to the Merovingian legacy.

The red curtain, a backdrop to the green altar cloth, droops down from its rail at the end, exposing a crucifix 'upturned'. As a supposedly hidden object, it sends a secret and questionable signal of its own. Tintern Abbey was a ruin and practically dysfunctional after the dissolution of the monasteries in 1536. The church of St. Michael's situated at the other end of the hamlet became even more important. The church stood on the banks of the river Wye which separated England and Wales. As positioned in the painting it could be described as a 'cradle,' nurturing the confusion which engulfed many of the decision-makers and aristocrats in the surrounding area. Loyalties were tested. In 1537 Henry VIII granted Tintern Abbey to Henry Earl of Worcester, the ancestor of the Duke of Beaufort.

As I have mentioned, I believe that Ann Bacon was buried in the South Porch of St. Michael's in Tintern. It was built during the Tudor period. Apart from any family connections, she played a very significant role in establishing the Protestant Church. As mentioned, she translated from the Latin "Apologie of the Anglican Church" by John Jewel, Bishop of Salisbury. The work was translated in order to reach a wider audience and was a justification for Protestantism in England. After the coronation of

Queen Elizabeth I in 1558, there was a re-establishment of the Church of England after of the death of Queen Mary, her predecessor and half-sister. The 'Thirty-nine Articles of Religion' were proclaimed by a Convocation of the Church in 1563. It was after this momentous proceeding that John Jewel's important thesis was published in 1564. Francis Bacon would have been three years old at the time, when Ann Bacon was Lady-in-Waiting to Queen Elizabeth.

Ann Bacon spent her last years at Mount St. Albans, known to have been lived in by Francis Bacon. It was situated next door to St Julian's near Newport nearby to Tintern. She died in 1610. St. Julian's was owned by the Herbert family, who also owned The Nurtons. It is my opinion that the inverted crucifix indicates that St. Michael's retained Roman Catholic sympathies, despite the burial of Ann Bacon and the established 'new order.' The Merovingian links were deeply rooted.

Returning to the figure of the Bishop of Gloucester and Worcester John Hooper, the artist seems to make clear St. Michaels' historical significance. The white wings of Mercury are a give-away, hinting at the Bristol 'White Friars,' founded by King Edward I. King Edward then is a key figure to understanding the story of the painting. He cut a formidable figure standing six feet two inches tall and was nicknamed 'long shanks.' Very much the patriarchal figure in his domestic life, he had three surviving children by his wife Eleanor of Castile. Having survived successfully several battles, including the 9th Crusade, he focused his last years on South Wales and in particular the Tintern area including Usk. I quote here a passage from my book 'The Hidden Chapter' published in 2011:[3]

> But in 1307 Edward I and Edward's daughter Joan d'Acres died. Joan had been born in Acres while her father was there on crusade in 1270.
>
> In 1290 she married Gilbert de Clare, the seventh in line of the powerful Earls of Gloucester and Hertford. But Edward insisted that the Earl hand over his lands to him and only returned them when Gilbert disinherited the children of his first marriage in favour of any offspring from Joan. This manipulation of inheritance laws was used several times by the king to suit his own interests. In this instance it meant that, after Gilbert's death in 1295, Joan became the sole owner of the estate of Tintern Parva, holding it for life – or as long as her father saw fit.
>
> Two years later Edward I decided to marry her to Amadeus V of Savoy only to discover that she had secretly married a young squire, Sir Ralph de Montherimer, a member of her first husband's household whom she had previously persuaded her father to knight. Edward I had a notoriously violent temper but in later life he learned to keep it in check. Joan must have pleaded well for her husband, for although de Montherimer was

imprisoned in Bristol Castle, a few months later he was released and all the Clare estates were restored to him and Joan. However, her death in April 1307 was followed shortly after by the death of Edward I in July. Her only son (Edward's grandson) Gilbert de Clare, inherited the Earldom and lands, including Tintern Parva. Seven years later, in 1314, he was cut down in savage, hand-to-hand combat at the Battle of Bannockburn and the fate of Tintern Parva came under review again, as we shall see in the next chapter. For the moment it is sufficient to note that Edward I had taken deliberate steps to ensure that Tintern Parva was in Royal hands.

In pursuing the changes of ownership of Tintern Parva at this time, I encountered another in the growing number of surprises that seemed to be accompanying my investigations.

I learned from Bradney that the customary inventory of land and property made after Joan's death include the church of St. Dennis at Llanishen, four miles from Tintern Parva. Here in a remote part of the Welsh countryside was a church dedicated to a saint inextricably linked with the Merovingian monarchy, whose fame and importance was due in the first place to the prominence given to him by the long-haired kings.

Originally, St. Denis (or Denys in old French spelling) was one of seven bishops sent from Rome to promote Christianity in Gaul in the third century. He became the first Bishop of Paris, but was persecuted by local pagans, imprisoned and beheaded. As with so many early saints extra-ordinary legends grew up around his death – one was that he managed to walk some distance from where he was attacked carrying his severed head. A small church was built on the site which over the years became a place of pilgrimage. This was later replaced by a Benedictine abbey built by the Merovingian King Dagobert in 630, who embellished it with gold and precious stones, none of which survives today. There is, however, some evidence of an earlier Merovingian interest in the site by Clovis after his conversion to Christianity in 496. Clovis's first wife, Evochilde was a pagan but his second child, Clothilde, was the Christian daughter of the King of Burgundy.

The small chapel at Llanishen was another reminder of the French dominance in this part of Wales. The distinguished genealogist Sir William Dugdale notes the existence of the church of St. Dionysius or Dennis alongside taxation returns for Tintern Abbey drawn up in 1291. That, as we know, was the year Acres fell and the year King Henry III's heart was transferred from Westminster Abbey to Fontevrault.

Many Merovingian priest-kings were educated and buried at the Abbey of St. Denis in Paris.[4]

Considering these Merovingian links that I see represented in the painting, I reminded myself of a significant find made in 1909 and quoted in my book "The Hidden Chapter" on page 79:

> ...the discovery of three triangular buckles at Caerwent and of one at Trostrey near Usk dating from the fourth century were proof of traffic between post-Roman Wales and Merovingian France. Trostrey was not far from Radegunda's chapel and provided a revealing piece in the jigsaw. Tewdric's possible Merovingian ancestry provided another.

King Edward I was buried in Westminster Abbey on 27 October 1307, just two weeks after the arrests in France of the Templars by representatives of King Philippe IV. Edward I's commitment and legacy in the Tintern area was evident for generations to come. His eldest daughter Eleanor had married Henry III Count de Bar in Bristol on 20 September 1293. They had three children, Edward de Bar, Eleanor de Bar and Jeanne de Bar. I mention them specifically by their full names because each of them takes their place in popular history unexpectedly. First, they were the grandchildren of King Edward I. Eleanor, named after her mother, married Llewelyn ap Gwain, the Lord of South Wales and therefore retained a power base in the region. But it was Edward and Jeanne, brother and sister, who became figures of intrigue. Both are included as consecutive 'Grand Masters' of the 'Prieure de Sion' or to use the official term 'nautonnier,' an old French term which means 'navigator' or 'helmsman.' Edward de Bar is listed as being such from 1307-36 (29 years). His sister Jeanne de Bar succeeded her brother from 1336-1351 (15years). Whatever their responsibilities were in the organisation, it was for a total of forty-four years. Their cousin Eleanor, the daughter of Joan of Acre together with husband, Hugh le Dispencer retained the holdings of Tintern Parva and Usk during the same period.

The list of 'Grand Masters' for this French organisation had first appeared in a document entitled the "Dossiers Secrets." The dates of those honoured began at 1188 and finished in 1918 when the artist Jean Cocteau took over. Research often throws up new and unlikely facts. My work on the painting seemed to generate a focus on the Tudor period with Merovingian undertones. The Prieure de Sion list of Grandmasters with 14th century direct links to Tintern for many years was a departure. But then I had to remind myself of another unlikely connection concerning the list some three hundred years later. Charles Radclyffe, Grandmaster from 1727-1746 until he was found guilty of treason along with many other Jacobites, was beheaded. His widowed mother was Mary Tudor, the 'base' daughter of King Charles II. She had married James Rooke, a military man from St. Briesal (near Tintern). Mary Tudor died in Paris on 5 November 1726 and shortly after her death Charles Radclyffe took over the Grandmastership of the Prieure de Sion in 1727.

The Merovingian pedigree claimed descent from ancient Troy, but certain Prieure documents trace it back to the Old Testament. For our purposes at present it is enough to remind ourselves of another nearby St. Michael's Church, Michel Troy, where Sir Philip Sydney's Latin tutor became Rector for a time, until 1611. Whatever the links between the two may be, the Prieure and the church, fact or fiction, Tintern and surrounding areas are part of it. The painting brings some of those facts together on this one canvas with the added caveat of Shakespeare's play "The Tempest," with "Richard III" thrown in for good measure. As for the real author of the plays, the painter gives a clear indication it is Francis Bacon, with more than a suggestion that the original manuscripts were buried with him in St. Michael's Church. So how does Bacon fit into this ongoing saga portrayed in the painting?

I have left one other question unanswered. The youth positioned at the top of the pillar, pointing with a short rod in the direction of the altar of St. Michael's I believe to be Theodore Dee, entered in the Collegiate Church records of Manchester (now Manchester Cathedral) as 'Tidder Dee'. He died on 9th April 1601 aged thirteen years as mentioned. He is the link tying Francis Bacon and the tragedy he lived to the canvas, a tragedy his son lived as well.

I have for some time believed that 'Tidder Dee' was the son of Francis Bacon. My belief was held before I had studied the painting under discussion here. This painting has confirmed my conviction. It has also resolved another perplexing mystery with which I have wrestled for some time. I have long recognised that Bacon had a 'double', the 3rd Earl of Rutland, Edward Manners as I have said. But when would Bacon need someone to act on his behalf? It had to be when he himself was unavailable for some length of time to perform his official duties.

I now believe that the hooded scryer, Edward Kelly, was wearing a disguise when seeking re-employment with John Dee. Having supposedly committed a felony (as he related to Dee), his ears had been removed and he was wearing a balaclava-type head gear for 'cosmetic' reasons – a subterfuge – very Bacon-like, very Shakespeare-like. He had by then already abandoned one pseudonym 'Edward Talbot,' having been sacked by Dee for that earlier deceit, passing himself off as Edward Talbot. This time Dee, recognising his undeniable abilities (and perhaps seeing through the disguise or perhaps unsure, or perhaps suspending his disbelief), decided to include him in his team for the European project being planned, on condition that he formally marry. A likely scenario…in my mind surely. Time passes. People change.

'Kelly' (Bacon) married Joan Cooper, but as recorded in Dee's diary, she seemed to be infertile. I believe that the young Jane Fromond, formerly lady-in-waiting in the court of Queen Elizabeth I, was attracted to Kelly and vice versa, contrary to some accepted belief. Of Jane Dee there are suggestions of a petulant personality recorded in Dee's diary entries when Kelly had confrontations with Dee. If Kelly was in fact Bacon, at some point he became desirous of an heir. I believe that having received instructions from the spiritual entity Madimi (via Kelly/Bacon as the scryer) a contract of 'cross-matching' marital behaviours was agreed upon and acted out. Theodore (Tidder) was born forty weeks later.

So, Francis Bacon had a son and heir. Jane Dee was Francis Bacon's son's mother. By this time, the stem of Jesse had become tainted by the drama surrounding the Tudor line of inheritance. Tidder Dee died on 9 April, that fateful recurring date.

The painting is a visual depiction of the tragedy surrounding the drama in this aspect of Tudor History and the artist knew it. It seems at this stage appropriate to quote Sonnet 152 in its entirety without further explanation:

> In loving thee thou know'st I am forsworn,
> But thou art twice forsworn, to me love swearing;
> In act thy bed-vow broke, and new faith torn,
> In vowing new hate after new love bearing.
> But why of two oaths' breach do I accuse thee,
> When I break twenty? I am perjur'd most;
> For all my vows are oaths but to misuse thee,
> And all my honest faith in thee is lost:
> For I have sworn deep oaths of thy deep
> Kindness, Oaths of thy love, thy truth, thy constancy;
> And, to enlighten thee, gave eyes to blindness,
> Or made them swear against the thing they see;
> For I have sworn thee fair; more perjur'd I,
> To swear against the truth so foul a lie!

The author (Bacon) felt trapped and was living a constant lie. His anguish is evident. The painting reflects in general an historical deceit. A repository of historical data must have been stored at The Nurtons. I believe that Oliver Lodge (following Sir Thomas Herbert) was the visual artist and chronicler of the accumulated data revealed in this 'allegory.' I have made every effort to trace the original painting without success and have attempted through all known avenues to seek the current owner.

Oliver Lodge was known to have moved from Tintern to Painswick and worked closely as an architect with Detmar Blow (1867-1939) the celebrated architect and designer. Oliver became the godfather to Blow's son in 1919. The family seat of the Blow family, 'Hilles House' is now used for "weddings and private hire." Current publicity indicates that the décor of the house prides itself on displaying a 'Mortlake Tapestry' in the decorations, which depicts Raphael's 'Acts of the Apostles.' The dimensions of Hilles House were designed to accommodate the tapestries. They take up a complete wall in the hall. Hilles House was built between 1914 and 1916.

The Mortlake Tapestry Industry was set up on John Dee's estate at Mortlake as a result of a proposal by King James I in 1619. For such a tapestry to be included in the original building at Hilles House between 1914 and 1916, when Oliver Lodge was working so closely with Detmar Blow whilst in residence at The Nurtons, is evidence of another unspoken truth!

CHAPTER TWELVE

The Nuremburg Allegory

ALLEGORIES IN ART, whether in visual media such as paintings, drawings, or plays upon the stage, or in stories told or even conveyed in jokes and jests are designed to convey hidden truths to the beholder, the listener. They are very like myths and legends passed down from generation to generation to uplift and edify those who were not there in the moment, engaged in the action of the time. We are all engaged ourselves in our own actions, in the conflicts and confusions and reality imposed by the 'present,' living our human condition. The eternal truths demonstrated by allegories in art and life illustrate and educate us as to how others have responded to the challenges of that reality and guide us to making our own choices when faced with crisis. I believe that's the purpose of Byrom's collection, passed down to us by Byrom and those who were engaged in 'hermetic' studies of the sixteenth and seventeenth centuries. I believe it's the purpose of all who have been engaged in Hermetic studies throughout human history. It makes me wonder if our world is but an allegory of a world more beautiful, more logical, more succinct than ours. (An observation by The Editor).

I now bring the reader's attention to another allegory in the visual media, another painting related to the hidden knowledge and understandings that these poets, artists, philosophers, scientists that I have for so long been investigating. I call it the 'Nuremburg Allegory.'

Upon the establishment of the wire and brass works in Tintern, experienced engineers were brought in from Nuremburg in Germany with skills to help train the locals. Some of these settled permanently in Tintern. The brass plates (catalogued as the Boyle Collection in the Science Museum) from which the theatre drawings in the Byrom collection were created were made in Tintern.

John Dee had spent ten days in Nuremberg in March 1589. Though the date and the days spent there are recorded in his diary, he does not say why he went or how he spent his time there. He and his family returned to England later in the year, having spent six years on the Continent, working professionally for the aristocracy with Edward Kelly.

The Nuremburg Allegory – The purchase

The Nuremburg Allegory – section in details

The Nuremburg Allegory – section in details

The Nuremburg Allegory – section in details

It is with this thought in mind that I introduce this allegorical painting that I call the 'Nuremburg Allegory,' which I have previously discussed in my book "The Hidden Chapter." At the time of that publication I had not been exposed to the Tintern allegory 'Stem of Jesse' (presented in the book "Landscape of Lies"). I was able to interpret the Tintern painting because of what I had already identified in the Nuremburg study. As will be seen, features in both support the interpretation of each and each other.

Unlike the Tintern allegory, I knew who the owners of the Nuremburg painting were and had seen the original. The painting had been bought by the Ezen family, Mayfair Restaurant owners, at an auction staged by Bonhams and Brooks in London (2001). The Ezens were refurbishing one of their restaurants at the time and their purchases were made with the aim of decorating it. This particular purchase however was thought by some family members that it would be of interest to a senior member of the family, Mrs. Ezen, an author of anthroposophical texts.

I knew Mrs. Ezen as Sylvia Francke. She, after having attended one of my London lectures, contacted me to discuss the interpretation of this unusual clearly allegorical painting. I received a set of photographs from her and became so intrigued that on my next visit to London I went to meet Mr. and Mrs. Ezen and inspected the original. I told them that I thought the content of the painting had something to do with Francis Bacon. They were pleased to co-operate in my developing interest and let me have copies of the purchase details and the origins of its placement in the auction. I was eager to learn more of the painting's provenance.

This painting was one of a number in "The Final Sale of Paintings from the Studio of Miguel Canals," Tuesday 4th December 2001, as advertised in the Bonham's auction catalogue. The catalogue goes on to give a little biographical background to Canals and his studio:[1]

Estudio Canals

Estudio Canals was founded by Miguel Canals who worked in the world of art from an early age. He began by helping his father, who owned a well-known artist's canvas factory frequented by painters such as Pablo Picasso and Joan Miro.

He graduated from the University of Terrassa in Barcelona as a Textile Engineer, going on to manage the family business. Towards the end of the 1970's Miguel Canals began to devote himself to what really fascinated him – painting. He created the "Estudio Canals" in Barcelona and soon the reproductions began to achieve worldwide recognition. The body of work was unusually characterised by its versatility in terms of both style and period, ranging from the 15th Century Dutch School to French

Impressionism.

Pictures from the studio have been exhibited worldwide with special exhibitions in Spain, Germany, Holland, England and the United States of America.

In the last few years Estudio Canals began exhibiting their own original works, the series titled "Birds and Fruit." These works were conceived by Miguel Canals.

Mrs Canals and her family have been running the studio since her husband passed away in 1995. This year will be the last annual sale of Paintings from the Studio as Mrs. Canals has decided to retire. The picture is listed as Number 36 in the catalogue and shown in a small illustration on page 11. On page 13, it is described as "After the German School, An Allegory, oil on canvas 35 3/8 x 23 5/8in (90 x 60cm).

I wondered how an artist's studio in Barcelona had come to put up for sale as a copy such a painting and wrote to the studio with an enquiry to that effect. I received no reply. At the bottom of the painting, in the style of German Gothic print, was a description of sorts. With the help of the German department at Manchester University to whom I am grateful, I was able to produce the translation:[2]

Well, then, there was an honourable man in great difficulties and distress on the water known as (Lake) Garda. He was afraid he would lose both his life and worldly goods. And when he reached the land from the water, he was rendered powerless by the stradiots from the Venetian mercenaries who had been searching for enemies everywhere. So being in even greater terror, he in his great distress with great humility called to the mother of God and to his twelve brothers here in Nuremberg and mercifully was helped.

Referring to the quote "twelve brothers here in Nuremberg" was a very helpful clue. But first Nuremberg was the hometown of Wenzel Jamnitzer, a goldsmith. A copy of his book on geometric figures had accompanied the brass plates when they arrived at the Science Museum. The book was published in 1568, the same year that Elizabeth I granted the charter for the brass works, also known as the wireworks in Tintern.

The "twelve brothers" referred to were elderly gentlemen, retired, who had worked in some specialised craft such as building or textiles, and lived together in an 'alms house' within the local community. To give some idea of the Nuremberg setting for this establishment I quote from "The Housebook" of the Mendel Twelve Brothers Institution:[3]

In 1510 a Brotherhouse and a Chapel dedicated to All Saints for 12 old men was founded by the metal and smelting master Matthaeus Landauer on the model of the Mendel Brotherhouse near the Chapel of the 12 Disciples at the cost of some 1450 florins. Just as Konrad Mendel had decided for his Brotherhouse, it was not allowed in this House, too, for a clergyman to be admitted, neither a priest nor member of an Order. At the beginning, just like the Mendel Brothers, the Brothers wore habits of different colours with a black border, a shallow felt hat and most had a beard. Depending on the circumstances at the time, the clothing varied. In the last hundred years of the Foundation's existence, the brothers wore a cloak which came down over their calves and a black three-cornered hat of felt. The grave of the Founder was found again during the renovation of the chapel after the destruction of two world wars.

The alms house, because of the earlier arrival of the Nuremburg engineers to Tintern, would have been familiar to the Tintern wireworks community.

I believe that Francis Bacon became a resident at the alms house in Nuremburg in 1626. It was an enclosed order, exclusive and self-sufficient, with skilled practitioners within the group of men. They would have provided the services and stimulus for Francis Bacon's scholastic and literary needs. Already sixty-six years of age, he would qualify for entry to such an establishment. His popularity in London had become diminished amongst factions of the aristocracy as a result of some sort of fraudulent activity or exposure regarding his legal career's clients and staff.

Bacon's death has been attributed to a chill, caught after he had conducted a biological investigation into a frozen chicken. The publication of that event was minimal, with missing facts. As already noted, his supposed death took place on 9 April 1626. The painting featured here is a compilation of three sections. I believe it represents three critical stages in Bacon's life, one of which was in Nuremburg after he had returned to Germany as Francis Bacon himself, an older man than he was previously when he had accompanied John Dee to that country, posing as Edward Kelly. The three stages in Bacon's life are bound together in the three sections of the painting.

A principle feature which must be considered is the boat; it is the most revealing. I include my interpretation of this significant feature from "The Hidden Chapter":[4]

Placed in the centre of the painting, it is drawn in such a simplified manner as to make us doubt its seaworthiness. The forecastle simply provides a platform for the youth to stand on. There is no wheel or tiller to steer this ship. Its fittings have been reduced to a minimum: a square sail, a mast and a few oars. It is meant simply to represent the *idea* of a boat. At the same time, however, its shape clearly resembles something else, something hinted at immediately above it in the allegory – a weaver's shuttle. As far back as the thirteenth century, in the *Zeno Narrative* the author writes, 'The Fishermen's boats are made like a weaver's shuttle.' A traditional weaver's shuttle was an oblong weaving implement curving inwards to a point at each end, on which the weft thread was shot across between the warp threads. In the painting, Joseph sits with his winding frame directly above the boat. The proximity of the two images is deliberate. What is the significance of a shuttle in this allegory?

The numerological equivalent for the word 'Shuttle' adds up to 100:

S H U T T L E

18 8 20 19 19 11 5

This is also the numerological total for 'Francis Bacon':

F R A N C I S   B A C O N
6 17 1 13 3 9 18 2 1 3 14 13

= 100 Total.

The word 'shuttle' is used only once in any play of 'Shakespeare.' It is spoken by Falstaff in 'The Merry Wives of Windsor' (Act V. Scene I):

> For in the shape of man, Master Brook, I fear not Goliath
> With a weaver's beam, because I know life is a shuttle.

Bacon (Shakespeare) inserted a code in the printed copy of his First Folio of plays. It is described as a 'sixth-line word-cypher,' in which Bacon inserted the numerological equivalent of his name into the sixth line up or down in the column of lines on each page. The same is true of entrances and exits and the beginning and end of each scene. Every column contained 66 lines, 66 being double the numerological equivalent of the name 'Bacon.' In the quote given here the cypher in question is the word 'shuttle.'

In this Nuremberg allegory I interpret the character with the winding frame (weaver's beam) spinning yarn as Jesus' father Joseph, positioned directly above the boat. Joseph was by biblical convention a carpenter, and the yarn (thread) drawn out links the figure to the 'beam.' In Act 1, Scene 5 of Shakespeare's Macbeth, there is another code which can be referred to. Lady Macbeth delivers a speech consisting of nine lines, some of which seem to be extraneous to the action. The first letter of each word on each line when strung together spell TAT, TAT, TAT, tatting being a word that means 'weaving' lace from a single thread of 'yarn' using a 'shuttle.' Obvious, Shakespeare (Bacon) was not!

I, having recognised that Francis Bacon is the main study of the painting, noticed that the colour scheme links other features of the tragic story together. I remind the reader that the Tudors considered themselves to be descended from the 'Stem of Jesse.' I had been fortunate enough by the time of my study of this 'allegory' to have become familiar with the works of several scholarly writers who had also concluded that Bacon was the son of Queen Elizabeth and Robert Dudley. I was saying nothing new. But the painting had additional features which seem to have escaped other's detection.

As with the Tintern allegory 'Stem of Jesse,' discussed in the last chapter, a colour scheme has been used by the artist to tie some of the 'threads' of the story together. The colour of the royal house of Tudor was green. The same shade of green is used in both allegories, highlighting the figure identified as Bacon in both. For example the youth standing at the aft of the shuttle (boat) and the green blanket strewn across the bed of the Virgin Queen (his mother), as well as one of the apocalyptic riders below whom I have identified as Bacon, are all the same colour green. The kneeling woman in 'widow's weeds' (dress code) provides an approximate identifiable date for this part of the story. After Bacon's reported death his widow, supposing him dead, would have been wearing such attire for a few days. We know that Bacon's officially recorded death date was 9 April 1626, so the painting would have been created after that time. 'Escorted' by his closest allies, he disappeared into the mists of time for his last years where he was, according to some scholars, creating other masterpieces for posterity in the frugal surroundings of the 'twelve brothers' house in Nuremberg.

Bacon's widow, Alice Barnham, whom he had married when she was fourteen in 1606, remarried Sir John Underhill eleven days after Bacon's reported death. Underhill was a member of the family household, a gentleman usher. Her marriage to Bacon had always been considered one of convenience. She was a wealthy woman in her own right and they had no children.

The shuttle scene recounts an occasion at Court when Bacon was a youth in his middle teens. It was included by author Alfred Dodd in his book "Francis Bacon's Personal Life Story" (published 1949). I include the event as described by Dodd:[5]

> The story continues that while Lady Scales, a Lady-in-Waiting, was repeating to her companions Cecil's malevolent whispers that Francis was the Queen's bastard son by Leicester, Elizabeth, overhearing the laughter from the adjoining room, dragged from the young Maid the reason for their merriment. By chance, Francis entered the room as the enraged Queen was violently beating the girl into insensibility. Attempting to intervene to save

her, he was then told of the scurrilous chatter that was going round the Court, and in the passion of the moment, the truth of his parentage slipped from the Queen's lips. In the violence of her anger she screamed 'You are my own born son, but because you have taken sides against your mother to champion a graceless wench, I bar you for ever from the Succession.'

Before the time of this incident, Bacon may or may not have known the truth about his own origin. Some students of Bacon maintain the discovery of the fact was the reason why Francis was sent with the household of Sir Amias Paulet to France when Paulet was appointed Ambassador in 1576. At the time he was a youth of fifteen. Before he returned to England he is known to have visited Rome and Venice.

The scenario seems to be a 'collective,' visually describing Bacon's appeal to his mother (nativity scene). The figure cowering centre boat would be Lady Scales. The figure to the right, with arms outstretched is Bacon's surrogate mother Lady Anne Bacon, protectively reaching out to her adopted son. So, what artist of the period would have been privy to the facts of Bacon's life sufficiently to create a visual summary in one piece? I repeat one sentence from the German narrative positioned at the bottom of the picture:

> So being in even greater terror he in his great distress with great humility called to the mother of God and to his twelve brothers here in Nuremberg and mercifully was helped

As well as providing the location of Nuremberg, the implication is that the painting itself emanates from Nuremberg.

I soon surmised that Joachim Von Sandrart was the most likely artist. Born in Frankfurt in the year 1606, the same year as Sir Thomas Herbert of Tintern, he was part of the 'De Bry' dynasty of artists, printers and publishers. As a student he had studied the rudiments of design and painting under Matthias Merian and Theodore De Bry in Nuremberg. Equally significant, he was invited to England by King Charles I along with the artist Gerard van Honthorst, whom the King had commissioned to copy paintings of historic significance. Sandrart eventually settled in Nuremberg. He became the director of the Academy there and married as his second wife Barbera Bloemaart, the daughter of a local magistrate. Around 1627 he is known to have met his cousin Michel le Blon and travelled with him to Florence and Rome. Le Blon's initials are written on some of the original drawings in the Byrom Collection as 'M L B.' Le Blon died on 14 October 1688. The De Bry's family businesses were well known in Europe as printers of intellectual and philosophical treatises as well as art books and maps. They moved within the circles of the Royal families and aristocrats and were very much respected for the quality of their work.

It was clear to me that the coat of arms positioned at the bottom left of the picture was part of the story. I identified it as being connected with the Brunswick Luneburg Royal family centred largely in Germany but with close connections in Britain and elsewhere. A significant feature of the coat of arms is the raised, single arm, identifying the design with Christian, a member of that family who had lost one arm in the 'Battle of Stadtlonn' in 1623. He died in 1626. Another member of this prestigious European family was Augustus, the Duke of Brunswick Luneburg. He was popularly referred to as the 'Code Man,' creator of the code 'Cryptomenylices et Cryptographiae libr IX,' 1624, under the name 'Selenus Gustavus' and for much of his life lived in London. He died during the Fire of London in 1666 aged eighty-seven.

Augustus was related by marriage to Elizabeth, King Charles I's sister. Elizabeth's daughter Sophia became the wife of Ernst Augustus. Augustus (code man) was his uncle. These close affiliations give credence to my understanding of the provenance of the painting. It is a fact that Queen Elizabeth of Bohemia, as King Charles I's sister was known, employed Sandrart's art colleague Honthorst as drawing master for her children. Both Elizabeth and Charles had grown up within the confines, protocol and culture of the Royal Court of King James I, their father, in London. The life and destiny of Sir Francis Bacon would have been well known to them. In all probability, Sandrart painted this 'Nuremberg Allegory' as a confidential record for the Brunswick Luneburg family. After all, it was Bacon's enforced anonymity and absence from the Succession which paved the way to their supremacy after the demise of Queen Elizabeth I.

* * *

Concerned as to the exact provenance of the painting, I wrote to the National Museum of Nuremberg and sent them a photo with some details concerning its sale and purchase. It was with some surprise that I received a response from a curator of the museum, Dr. Daniel Hess, an excerpt from which I reproduce here:[6]

> The original of your canvas painting is still exhibited in our collection. It is a panel painting, attributed to Paul Lautensack, from 1511 (Inv. Gm 196) which shows the Nuremberg patrician and merchant Stephen I. Drawn in distress at the Lake Garda in Northern Italy. This is one of the most interesting votive-pictures in early German 16th century paintings. It hung once in the Chapel of the Mendelsches Zwolfhruderhaus (Mendel Twelve Brothers House) not in the Landauer Zwolfhruderhaus where Durer's

Famous All-Saints-panel was situated. For more information about our Lautenstock painting, please look up our catalogue of the paintings of the 16th century in the Germanisches Nationalmuseum by Kurt Locher, Stuttgart 1997 S.306-309. I would be very happy if you could send me a foto and the documentation of your copy to complete my documentation on the original.

13, September, 2002.

Over the next months, photographs of the two paintings were exchanged, the German Museum copy was studied and every effort was made to obtain information which would satisfy the serious scholar, together with the owners and myself, of its provenance.

At the outset I could not accept the suggested date or description of the museum painting as accurate in the original report, considering the story which this 'Allegory' obviously told. The painting is too vivid and precise in its composition when one understands what that story actually is. In addition, placing the photographs side by side, certain details differ. For instance, the apocalyptic rider dressed in white has black ringlets in the Museum's version, a detail that mattered to the artist. It was very clearly of significance and not a fashion note. Also, the fact that the German explanatory note is positioned outside the main frame in the museum piece, whilst in the Ezen 'copy' all is within the single ornate frame. It must also be said that from the Museum Catalogue it is known that there was a copy made of the original. I quote that passage from the report: [7]

> A copy represents the place of the votive picture in the art cabinet of the Praun family. It was named for the first and only time in the inventory of 1719 (as) number 255, 'copy of a tablet which used to hang in the Chapel of the Twelve Brothers by the Charterhouse, on which Stephen Praun's first solemn promise was painted on wood 2 feet 9 inches high, 1 foot 9 inches wide.' But by 1878 the painting could no longer be traced.

I noticed that the documentation provided by the museum contained information which seemed to differ from expert to expert within the report itself. Certainly, my understanding did not match up to the references made available. At the time, I felt that more factual certainties were required. I included a summary of my interpretation in "The Hidden Chapter" referencing the Nuremberg entry for any interested reader or researcher.

While understanding that Miguel Canals was a professional copyist of the paintings of others, I was uncertain as to which of the two was the original and which the copy because the date 1511 held no meaning for me in my understanding of the content. I understand that the frames could have been changed over the years and the actual artist might be identified by the techniques employed in its creation by experienced experts in the field. However, I have not sought to have that done.

I felt at the time that the story of the 'Allegory,' as portrayed in it, is such that the painting could create or result in such confusion that the matter might be dismissed entirely. I was not prepared to compromise the data that I had collated and collected from it. There was too much of a gap between the Nuremberg Museum's description of their version and the Ezen's 'copy' as I interpreted it, to try to settle the matter with finality. It was not until many years had passed and the Tintern allegory 'Stem of Jesse' had been brought to my attention by Rev. Mr. Cryer that some questions could be considered afresh. There were aspects of both which I saw had connections to each other. The 'Nuremberg Allegory' had arrived at the Bonham Auction House in London from a professional artist's studio in Barcelona, Spain. At the time of my earlier enquiries I could find no answer from that venue. I had drawn a blank.

* * *

Having accumulated long-lost forgotten facts regarding Tintern's history, some of which have found their way into my books already, I considered the next step to be to obtain any information in the public domain concerning the 'Prieure de Sion.' The 'Prieure' had deep connections to Tintern. While I was well-aware of the disputed claims and misinformation that has been generated in recent years about the organisation, I needed to find out more. Trusting certain accounts and records I used, I became particularly interested in an article published in the Journal 'Vaincre no. 3' dated September 1989, and reprinted from the "Priory of Sion Archives of Paul Smith." The article is headed "Some Archives of the 'Priory of Sion' discovered in Barcelona." The first paragraph I quote:

> Since the end of 1939, certain archives of the 'Priory of Sion' have been located in Barcelona, where the Comte de Saint Hillier deposited them in anticipation of the events that were to lead to the Second World War

Given that my own documentation had already proven 'Priory' associations of Tintern with some of the early Grandmasters listed in the 'Dossiers Secrets,' as published by Baigent, Leigh and Lincoln, I was intrigued by this article's edited list of Grandmasters. I include the complete list from the article:[8]

The Grand Masters of the PRIORY OF SION:
1681 ....Jean-Tim. NEGRI d'ABLES
1703 ....Francois d'HAUTPOUL

1726 ....Andre Hercule de ROSSET
1766 ....Charles de LORRAINE
1780 ....Maximilien de LORRAINE*
         (period of the French Revolution)
1801 ....Charles NODIER
1844 ....Victor HUGO
1885 ....Claude DEBUSSY
1918 ....Jean COCTEAU
1963 ....Francois BALPHAGON
1969 ....John DRICK
1981 ....Pierre PLANTARD de St CLAIR
1984 ....Philippe de CHERISEY
1985 ....Patrice PELAT
1989 ....Pierre PLANTARD de St CLAIR
1989 ....Thomas PLANTARD de St CLAIR

Of special interest was the year that the list begins. 1681 was the year that Sir Thomas Herbert of Tintern died. I had for some time considered that it was his understanding and knowledge that formed the structure of the Tintern allegory. He, as we know, was the owner of The Nurtons, which I knew had historical links to the 'Priory' organisation. The building in Nuremburg occupied by the 'Twelve Brothers' had disappeared during World War II.

The Barcelona Priory Article gave me much to consider. Had the Ezen copy been part of the deposited 'Priory' Archive of 1939? Was the Alms house connected in some way with the Priory? By that date Jean Cocteau was the recorded Grandmaster. It is known that he spent some months in the City of Girona, approximately thirty-eight minutes train journey from Barcelona. The author Patrice Chaplin in her book "City of Secrets" describes meeting Cocteau there in 1955. She took part in a film he was making. She also recounts how her curiosity drove her to peer through the windows of a deserted building behind Girona Cathedral where she saw Cocteau taking part in some sort of ceremonial ritual.

Girona Cathedral claims to possess a throne used by Charlemagne. A painting by Durer, with Charlemagne as the central figure, was hanging in one of the two 'Brothers Houses' in Nuremberg until Rudolf II, the Holy Roman Emperor, had it transferred to his Palace in 1585. Charlemagne had married a Merovingian princess, linking himself to the 'royal blood' descended from Jesse. John Dee had worked closely with Rudolf II and had himself visited Nuremberg in 1589, shortly after the painting was moved. I took these interconnections seriously, especially since Cocteau had such a close connection with Tintern through Sir Percy Loraine, a relative of John Baldwin, who was buried in St. Michael's Churchyard in

Tintern. The Barcelona Priory document, though of interest in general, is not relevant to my search for accurate data, except for the following quote:

> The origins of the 'Priory of Sion' are actually quite modest.
> The Priory stems from Razes and is only a more or less direct
> successor of the Children of St. Vincent.

That detail will be dealt with more later. Note the name Razes.

And what of Comte de Saint Hillier, the depositor of the Priory documents for safe keeping in Barcelona just before WWII? Born as Bernard Saint-Hillier in 1911, he died in 2004 aged ninety-two years. Apart from any other offices he held I found it interesting that in 1968 he was promoted to General de corps d'armie, commanding the 3rd Military Region in Rennes and sitting at the Counsell superior de la Guerre. Rennes-le-Chateau, central to the Priory story as it is currently understood, would be part of that landscape.

Returning to the 'Nuremberg Allegory' and its provenance, I believe the disappearance of one of the two paintings, either the original or the copy under discussion here, from the city of Nuremberg in 1878 is linked to the Priory organisation and their interests at the time. Victor Hugo was Grandmaster in 1878 and was known to have had contacts in Tintern as we shall see. Whether the Ezen copy is in fact the original is a matter for expert scrutiny. But the substance remains. The story represents an account of the uncrowned Tudor King, Francis Bacon.

CHAPTER THIRTEEN

Disclosures – Released

For quite some time I resisted the temptation to discuss the Shakespearean legacy of authorship and the deceptions surrounding Francis Bacon's life as I believed it to be. I came to refuse invitations to give lectures on my published books and turned down the possibility of focussed documentary films. The Shakespeare industry was flourishing in the U.K. and abroad. Any contribution that I might make would receive a cacophony of displeasure and doubtless 'mockeries' from the already established perceptions of the academic fraternity, if not others.

But it was the distinguished Freemason and author, the Reverend Neville Barker Cryer, who had given me the initial 'tool' to investigate what I consider to be the 'tell-all' painting 'Stem of Jesse'. Although he had died before I chose to study it seriously, over the years we had many discussions about historical information not easily available to the general public. The Byrom Collection had opened several doors for me in my studies and sometimes the question of relevance and suitability would be a matter of debate. Our meetings were always conducted in a semi-formal way – there was no time-wasting. On one such occasion I was intrigued by his reaction to my growing interest and concerns regarding Francis Bacon's place in world history. His answer surprised me. He said with absolute conviction, as was his style, that my responsibility as a researcher (and chronicler) was to report my findings and it was up to others to decide what to do with them. My reaction was one of hesitancy. My sense of responsibility resulted in me becoming cautious and rather withdrawn. I had already some years ago had to deal with unwelcome behaviour regarding the design of the new 'Globe' building in London.

I have now to present another visual 'allegory' pertaining to my research. It came into my possession towards the end of September 2013. The relevance will become clear as I proceed. The illustration, an 'original,' was put up for sale on Ebay. It again was purchased by the Ezen family who then sold it to me. The timing of the sale seemed pertinent. I had recently been interviewed by Jamie Theakston for a TV programme about "The Byrom Collection" and the Globe theatre design. It was for a series called 'Forbidden History' and was the first public interview I had given about the subject. My interpretation of this 'new' illustration became a major undertaking for me shortly thereafter. As I learned more about it, the facts when placed alongside my other research over the years forced me to become even more reclusive for a time. There was much accumulated background material to consider.

This allegorical illustration had been advertised as 'Strange 1804 Paris Holy Grail Masonic Templars Occult Coded Manuscript.' The sale was advertised for Saturday 28 September 2013[1], the evening before the first 'Forbidden History, Episode 1' was shown on TV. Mine was Episode 6. The sale was concluded the following day. I later came to believe that whoever put that item up for sale knew about the forthcoming series beforehand. The episode in which I was featured was shown on 1st November, five weeks later, although I had been featured many times in the advertisements leading up to the launch of the show. Included with the sale papers was a copy of the 'bidders account,' which I subsequently obtained with the illustration. The numerical values of the bidding prices echoed historical events and personalities, some with Masonic associations. It seemed to me to be a 'rigged' sale with concealed information for the purchaser, when and if it might be understood.

To give a description of the illustration: It is drawn on a page taken out of a very large Bible. The page was blank on one side originally and on the other side is a printed illustration of 'Christ feeding the multitude.' The heading to the printed side carries the statement "Engraved for the Reverend Doctor Southwell's Family Bible." The size of the page is fifteen and a half inches by nine and a half inches. The blank side is now filled with various scribblings of imaginative flair by an artist who it seems had much to say. In reproducing it here the size has had to be critically reduced. Nevertheless, the story should be told for the sake of the artist and the information he wished to convey. Clearly, I bear personal responsibility for my own interpretation of his genius and skill.

But first I must draw attention to the fact that approximately one third of the left side of the page is stained, as if with blood. Complex coded writing, handwritten in ink fills much of the space, headed by the words:

PARIS 1804. 7 Rue de Guillotine

The artist reminds us that on 25 June 1804, the dedicated counter-revolutionary Georges Cadoudal was guillotined in Paris, with eleven of his brothers-in-arms. They were supporters of the 'Bourbon Royalist' dynasty, which had been supplanted during the French Revolution. Cadoudal had taken refuge in England after being imprisoned for his activities and then escaped on several occasions. He was backed by supporters in Britain. Later that year on 2 December 1804 Napoleon was crowned Emperor of France at Notre Dame Cathedral.

'1804' disclosures examined

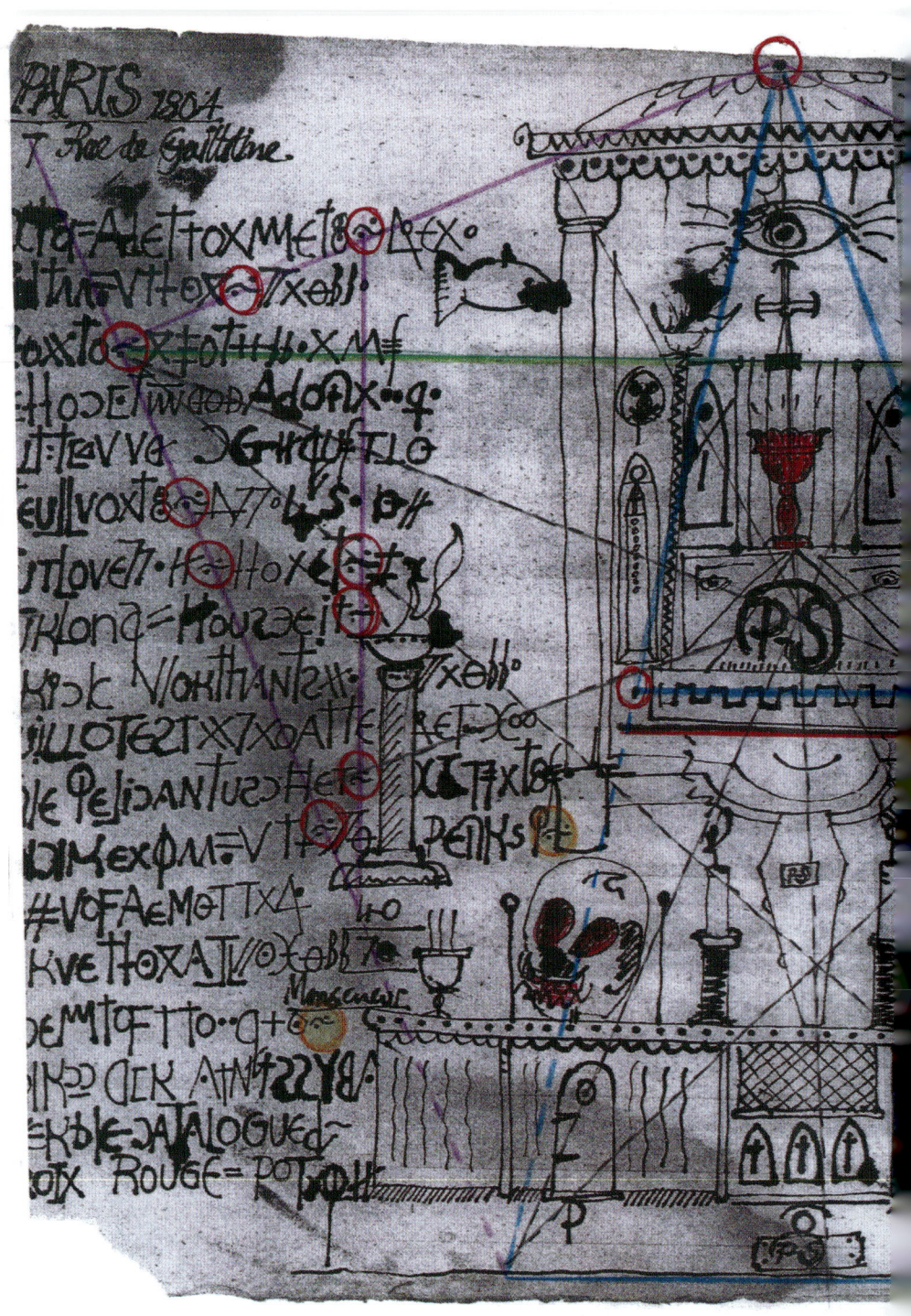

The coded geometry of '1804'

Before being executed, Cadoudal is said to have shouted to the crowd "And now, it's time to show to the Parisians how Christians, Royalists and Bretons die." That statement sets the tone of the illustration presented here and accounts for its construction. It soon became clear to me that it is a visual chronicle of a period in history, using the altar of the French Church in London, Notre Dame de France, as its centre piece. Jean Cocteau's commission to create the artwork at the church was undertaken between 3 and 11 November 1959.

I had already featured the same altar in my book "The Hidden Chapter," so I recognised it immediately. The secret information concealed within the tapestry there, involved with Cocteau's church-work, was very disturbing to me at the time. The central symbol strategically placed in this 'new' illustration was yet another beginning. The 'PS' located within the coiled inverted C at the centre stands for the 'Prieure de Sion,' the secret society encountered in the last chapter.

For the next few months, I became locked in the unfolding of a mystery surrounding the sale of this 'piece' and the secrets locked up in it waiting to be discovered and understood. The more I came to recognise in the details of its presentation, the more intrigued I became. For a time, I considered whether the illustration was some sort of hoax or even a joke. The Prieure de Sion had become a focus for false information of sorts over the years. But this illustration is, in my opinion, a work created by 'someone' at the same time that Cocteau was working on the mural in the English church of Notre Dame in November 1959. What I had acquired was an original and I believe the artist could only be Jean Cocteau himself.

Cocteau became the Grandmaster of the Prieure in 1918. In the mural he includes a depiction of himself visually to the left of the altar. In this illustration, he is there in the same position, not as a physical likeness but as the figure 70. He was seventy years old in 1959. And then the number 40. He had been Grandmaster of the Prieure de Sion for forty years by then.

The intrigue deepened when I recognised that the structure is based on Pythagorean geometry. But even further, I realized that it is based on the '72 measure' I discovered in the Byrom collection of drawings. The latest of those was created in 1732. To understand this, one has to identify the dot at the top centre above the canopy holding the two pillars together above the altar. The geometric features emanate from that point. Numerology is one of the features of this piece, used in a very purposeful way. Whoever created this illustration used the '72 measure' to create features within it to tell a story. I had created such a ruler by collating and comparing the size of the pieces of card and paper in the Byrom collection previously. The artist must have known the same system exactly. I found this very interesting.

On 2 December 1804, Napoleon Bonaparte was crowned Emperor of France. His obsession with the Merovingian genealogy was apparent at the ceremony. He had three hundred miniature bees made of solid gold fastened onto his coronation robes. They had been found in the tomb of the Merovingian King Childeric I, the son of Merovee and father of Clovis, when it was discovered in the Ardennes. Merovee is himself featured in the painting discussed in a previous chapter. The bee is a sacred symbol of the Merovingians.

The letters 'PS' are featured so many times in this picture that the artist is making it clear that the 'society' is heavily represented in the story that is told. Above the central 'PS' is a vessel coloured red. I interpret this feature as the Holy Grail, flanked on either side by sentinels. The 'society' is protecting what it stands for. The red line underpinning this section is depicting the Merovingian bloodline, linking it in some way with the red vessel.

Napoleon, in his determination to gain an understanding of this bloodline, commissioned the Abbe Pichon to compile genealogies. Wishing to associate himself with this legendary group of priest kings, he married Josephine, whose daughter and son (Bonaparte III) from a previous marriage carried on that 'royal blood.' The genealogies created by Pichon are said to form the basis for Prieure documentation thereafter. Napoleon, in marrying Josephine, adopted the two children from her previous marriage Alexandre Beauhernais, a Merovingian descendant. The two children would therefore be of royal blood as 'Bonapartes.'

In studying the illustration, I consider Abbe Pichon, the creator of the genealogy, to be standing on the goblet at the end of the right-hand side of the altar. The two heads with red colourings I think must be the two adopted children of Napoleon. Napoleon himself I identify as the figure bottom right with the famous curls in his hair. I had by this time spent a considerable amount of time studying the technique and detail within this complex piece. Despite the ink splodges and sketchy appearance of the characters, the details I thought were very accurate and skilfully produced. I decided at that point that it is the work of Cocteau himself.

I had known much of the background of the organisation called 'The Prieure de Sion' for some years before I acquired this piece. I had been sent a copy of "The Holy Blood and The Holy Grail," authored by Baigent, Leigh, and Lincoln, by the late Sir Miles Huntington Whiteley, the grandson of Prime Minister Stanley Baldwin. Whiteley was very interested in my research and recognised some sort of connection, although he knew not what at the time. Recognising that one of the earlier grandmasters listed in the book, Charles Radclyffe was a friend of Byrom, I attended a conference at Shugborough which was organised for two of the authors,

Michael Baigent and Richard Leigh. We formed our own discussion group afterwards in order to place Byrom historically within the network of the mystery.

At the time 'The Three Holies,' as Baigent, Leigh and Lincoln were popularly referred to in the trade, were preparing their next book "The Messianic Legacy." I had regular meetings with Richard Leigh in London and an arrangement was made for Michael Baigant to study the original drawings of the Byrom collection with me here in Manchester. It was during one of those meetings that Richard Leigh mentioned what has turned out to be a very significant fact years later, especially while studying the 'allegory' under discussion here. He told me of a meeting that the group of three had had with 'M. Plantard,' the acknowledged leader at the time of the 'Prieure de Sion' on 17 May 1983.

Plantard spoke of four parchments, three of which were genealogies said to have been found by the French priest Sauniere in pillars of his church at Renne-le-Chateau during renovations there. Plantard produced those documents at their meeting. The parchments had allegedly been purchased by the 'International League of Antiquarian Booksellers.' On the first page of the documents shown to the authors, Richard described to me what was to be published in their forthcoming book about the meeting. I quote the relevant passage from "The Messianic Legacy":[2]

> The first Authorisation to the Consular-General of France: In the ensuing text, three Englishmen were cited; page in the documents shown us by M. Plantard was headed 'Request for the Right Honourable Viscount Leathers, CH, born 21st November 1883 in London; Major Hugh Murchison Clowes DSO, born 27th April 1885 in London; and Captain Ronald Stransmore, OBE, MC, born 3rd March 1888 in London. These three gentlemen requested permission from the Consulate-General of France to export from that country: three parchments whose value cannot be calculated confided to us for purposes of historical research'...

The intention was to investigate the parchments and to demand the recognition of "Merovingian rights made in 1955 and 1956 by Sir Alexander Aikman, Sir John Montague Brocklebank, Major Hugh Murchison Clowes and nineteen other men in the office of P.F.J. Freeman Notary by Royal Appointment." The arrangements for the handing over and dealing with the parchments was very complex and is described in some detail in "The Messianic Legacy." At one point another quote mentions:[3]

> Lord Selborne declared that the documents in question, deposited by Captain Nutting, Major Clowes and Viscount Leathers at the International

League of Antiquarian Booksellers, 39 Great Russel Street, London would be placed 'on this day' in a strong box of Lloyds Bank Europe Limited. No divulgence of them was to be made. At the bottom of the page there was Lord Selborne's signature.

I know from the paperwork associated with this that the date was 29 August 1956. One of the three Englishmen involved in this procedure died on 13 December 1956. This was 'Major Clowes.' His full name was Major Hugh Murchison Clowes, one of the three who provided another strong link between the 'Prieure de Sion' and John Byrom himself. Byrom's own diary talks of meetings with Charles Radclyffe, another listed Grandmaster. But the Clowes family was very prominent in Manchester and Byrom had very close connections with it. When in London Byrom would often stay with his cousin Joseph Clowes and on one occasion in June 1729, having left his chambers, he went straight to meet Charles Radclyffe at the 'Sword and Blade' Inn.

After re-examining some of my earlier research it became evident that Hugh Murchison Clowes was a direct descendant of the Manchester Clowes Estate family. This had far-reaching consequences. I had studied the responsibilities of the Clowes Estate in Manchester. Now converted, it once housed the Jacobite Thomas Syddall who was executed around the same time as Charles Radclyffe, the Priory of Sion Grandmaster. In my earlier books I discussed elements of those historic findings. Hugh Murchison Clowes became more than a curious individual to me. I traced and visited the current 'Head' of the family who at the time lived in Derbyshire. He was able to put me in touch with other members of the family, one of whom was William Beaufoy Clowes, who had written a book entitled "Family Business."

We began a correspondence which culminated in a meeting with him at his home in Northamptonshire. I was anxious to learn more of the unwritten background of this Hugh Murchison Clowes. The Clowes family was a distinguished family of printers for well over two hundred years, printing the catalogue for the British Library and specialising in the printing of Bibles. It was also responsible for printing material for the Government of the day. Hugh Murchison Clowes joined the Board of Directors in 1935. By profession he was a solicitor and worked for 'Hunters' in Lincoln's Inn, London. I had an interview with a representative of that firm, Mr. Bishop, together with the Beadle, Mr. Hope in February 1987.

I had already been in correspondence with Alan Philpotts who was 'articled' to Hugh Clowes. He told me that Clowes had been a partner in the firm and was Clerk to the Masons Company. He was also a part-time director of a shipping firm, 'C-Hayes Wharf,' but "solicitor was his main occupation." Unfortunately, Philpotts had died ten days before my meeting at 'Hunters,' but Mr. Bishop and Mr. Hope were very eager to be of help and had in hand ready to show me, a picture of 'J. Byrom' in the office. Byrom was 'a benefactor of the firm' which had been started in 1715 by Ambrose Newton. A dinner had been held each year in Byrom's honour for some time. But another interesting fact emerged during this meeting. Working with Clowes was another long-serving solicitor, James Louis Theodore Guise. He was born in Calcutta, Bengal, India in 1910, went to Oxford University and joined Hunters in 1941. The family of 'Guise' was very much caught up in France's Royal family politics along with the family of 'Lorraine.'

It is difficult to imagine that the Hunters employee I spoke with would not have been aware of the connection. Guise would have been forty-six years of age at the time of Clowes's signing of the paperwork in connection with the Merovingian genealogies in August 1956. Clowes death aged seventy-one three months later stopped further facts from being checked by me. His death certificate describes him as a "Company Director of a Shipping Company." The fact that he was involved at such a level in investigations concerning the Merovingian legacy was enough for me at the time.

"The Messianic Legacy" was published in 1986. 'The Three Holies' branched off developing their own researches, collectively and individually. I focused on investigating Byrom's collection of geometric drawings, especially in connection with the Elizabethan theatre designs and then the brass plates associated with them.

With those years behind me, I now return to the allegorical illustration and my interpretation of it. I noticed that there is a code on which the structure is based. The 'dot' at the top is a focal point to start, as I have said. Below, where the candles are standing is what I interpret to be an altar. The dots on the altar rail must be counted. There are thirty-nine dots, no more, no less. A deliberate number. What do they represent? Cocteau would have been well-aware of the logistics regarding security and confidentiality surrounding the transfer of parchments declaring Merovingian rights in 1956. That business took place at '39 Great Russell Street.' It was only three years later in 1959 that he was carrying out his mural mission at the French church in London.

Hugh Murchison Clowes was part of that arrangement as well. Cocteau is putting a 'marker' down in this drawing indicating that it was an event formally arranged in a responsible way, via 39 Great Russell Street. The year 1959 was years before Baigent, Leigh and Lincoln's books recorded the events. The first edition of "The Holy Blood and The Holy Grail" was published in 1982, twenty-three years later. The 'allegory' in this case seems to back up their account of events. Cocteau was Grandmaster of the 'Priory of Sion' at the time.

* * *

At this point, my enquiries took me to a more serious turn. The page in question had been extracted from a large Bible. I had noted the name 'Reverend Doctor Southwell's family Bible' on the reverse side. As I said, this 'allegory' was sold on 'Ebay'. It was 'Ebay' that provided me with another clue. The company advertised another similar Bible for sale as follows:

> 1773, Southwell Bible Apochrypha Folio 100 Engravings – Antiquarian and Collectible £570 – In stock. Undated Large volume (16 inches X 10 1/2 inches wide by 4 and half inches thick) fully collated and complete Henry Southwell Universal Family Bible.

The size is the same as the Bible from which the page I describe above was taken. That was in 2015, two years after I had purchased my own 'page.' The size of the Bible confirmed it was taken from an identical one as my purchased page. The sale price, apart from anything else implies that the removal of the page from such a solemn artefact was a deliberate act of callousness. Or perhaps even reverence.

Who was 'Henry Southwell?' Born in 1729 and died in 1779, he was an Anglican Reverend who published a book on Christian martyrs, which focused on the injustices endured by Christians through the centuries worldwide. He was the son of Edward Southwell (d. 1748) and Jane Dymoke (d. 1761). Remember that John Dee was visited at Mortlake by members of the same Dymoke family. Regarded as a fanatic in his day, Henry Southwell's father had died when he was a young man and it seems that his mother was a formidable influence on him. She was a member of the Lincolnshire Scrivelsby Manor family of Dymokes, the holders of the office of 'King's Champion,' if the reader recalls what it means. It seems that Jane was named after an 'Aunt Jane' whose husband Hon. Charles Dymoke had acted as 'Champion' at the Coronation of the Protestant King William and Mary in 1689.

I had already learned of the execution by guillotine of Cadoudal along with his eleven brothers-in-arms in 1804. It seems that Southwell's book, a study of Christian martyrs, was in some way connected to the choice of using that specific page from his published Bible. Who owned that Bible, now minus one page remains an open question. I wrote to officials of the French Church in London and they were unable to help. Since Cocteau is known to have had two assistants helping him with the altar mural, I think it quite possible that he was illustrating this Bible page during the same week.

It is now time to turn our attention to the inverted C surrounding the PS resting above the platform, above the altar confirming the identity of 'The Prieure de Sion.' The PS appears four times altogether in the illustration, three times on what I describe as the central column, including the vertical coffin. The coffin is about to disappear into the 'Prieure' protected zone of the Organisation (in this illustration). The fourth fish figure is encased within the frame of the 'silenced' or 'mute' fish, bottom-right; there are five fish in the illustration with the PS attached. Each of the fish (apart from one) has an appendage, a message or signal to be interpreted and understood. The message then must be placed within the context of the whole. The only fish that has no addition is the one at the top, to the right of the picture, a Christian symbol itself.

The French word for fish is 'poisson.' I suggest that the fish drawings in this instance of the 'allegory' stands for 'Poussin', the artist and painter Nicholas Poussin as well as the Christian symbol, a double meaning. I knew that Poussin was responsible as artist for one of three reproductions of paintings that Berenger Sauniere had bought while on a visit to Paris, before returning to his village of Rennes-le-Chateau where work was continuing on refurbishing his church. Poussin's painting is entitled 'Les Bergers d'Arcadia' or 'The Shepherds of Paradise.' King Louis XIV became the owner of the original Poussin painting, kept in his private apartments in Versailles. The subject of the painting is three shepherds and a shepherdess whose attentions are focused on a large old tomb which bears the inscription 'Et in Arcadia Ego,'or 'And I in Paradise.' (Heaven.) Some observers have thought there is something missing from this statement, reading it instead "And in heaven I...," not knowing that word order is not important in Latin. The statement is complete.

The story of 'Les Bergers' and Poussin have created an intrigue which has been occupying the curious and investigative mind for centuries. The illustration under review here appears to be including elements of that mystery. As early as 1656 Poussin, who lived in Rome, received a visitor. He was Abbe Louis Fouquet, the brother of Nicholas Fouquet who was the Superintendent of Finance to Louis XIV, King of France, who did not own the painting at the time. Abbe Louis wrote to his brother Nicholas, telling him of the meeting with Poussin and of secrets relayed to him by the Artist. That letter eventually became part of the archives of the Cosse-Brissac family, a prominent French Freemasonry family since the eighteenth century. A quote from the letter is said to be as follows: [4]

> And what is more, these are things so difficult to discover that nothing now on this earth can prove of better fortune nor be their equal.

Having received the letter from his brother, Nicholas Fouquet's correspondence was confiscated and he was placed under house arrest for the rest of his life. He was put on trial for other offences to disguise the real motive. In the 1970's an identical tomb to that depicted in the painting was located a few miles from Rennes-le-Chateau and the nearby Blancheforte estate. But even nearer, in the churchyard of the Rennes-le-Chateau church, Sauniere exhumed a flagstone with curious carvings on it. It dated from the seventh or eighth century. There was also a headstone and another flagstone, marking the grave of Marie, Marquise d'Hautpoul de Blancheforte. This monument had been erected by Sauniere's predecessor Abbe Antoine Bigou who had been involved in some way in creating some of the parchments that had been discovered earlier. This grave and the writings on it have caused confusion and controversy ever since.

There was a considerable amount of activity taking place at the time at Renne-le-Chateau, including a new vicarage for Sauniere named 'Villa Bethania.' The circumstances surrounding the changes in the immediate area have been very well explored by The Three Holies and others, and I am very grateful to them for enabling me to focus on what has to be a striking fact in this allegorical illustration.

I learned from the details describing Marie de Blancheforte's headstone, dated "17th January, 1781" that there was an exact replica inscribed on it of the inverted C with the 'PS' inside of it, as it is in this allegorical illustration's central column. It is also a fact that the same symbol is used as a signature on the first parchment found in the pillar supporting the altar of the church in Rennes-le-Chateau in 1891. I had seen the symbol years later in publications dating from the 1980's.

Cocteau, as Grandmaster from 1918 would have had privileged knowledge regarding the 'Prieure-de-Sion' and known its significance. He chose to use them, not on the mural in the French church, but in this 'allegory' where he provides additional information in symbols and code. Symbolically, Poussin and Fouquet are included in the picture as well, together with loyal members of the Fouquet family. Nicholas Fouquet was placed in prison in 1665, as I have mentioned. That year Poussin died in Rome. Despite the efforts of Louis XIV, it took another twenty years to acquire the elusive original painting for his private collection. Nicholas Fouquet had died four years earlier in 1681.

Marie de Blancheforte was a direct descendent of Bertrand de Blancheforte, who had succeeded as fourth Grandmaster of the Knights Templar in 1153. He is credited as being one of the most influential of Templars, transforming them into a masterful and influential organisation. His estate was not far from Rennes-le-Chateau. The next year in 1154, Henry Anjou became King Henry II of England. The lords of Anjou, the

Plantagenet family, are part of the genetic lineage of the Merovingian heritage. 1154 is recorded on the enigmatic triangular drawing in the Byrom collection, which I have associated with historic dates and Westminster Abbey in my previous books.

Using my personal '72 ruler' created from that collection, I noted that the height of the two pillars, from the plinth to the top measure '72.' A (fish) poisson is positioned on either side at the top. The one on the left with an ink splodge covering its mouth suggests that Poussin had been silenced by the Templar organisation's 'in-house' group, the 'Prieure de Sion.' It is apparently intended to protect what is represented between the two pillars central to the picture.

This 'enclosure' receives the vertical coffin, which also contains the letters 'PS' on its surface lid. When studied alongside the mural painted in the French church, it sends out a signal which I have no answer to. In the same position in the French church, Cocteau's mural depicts the Crucifixion as described in the New Testament. In that depiction, the bottom half of Jesus's body, his legs and feet, are supported by a weeping Mary and others. In this '1804 Allegory,' as I have christened it because of the date at the top, the scenario replaces the legs and feet with a vertical coffin and the 'PS' letters. The artist had a reason for displaying it this way.

By this stage I had learned that the French Church archives were closed for enquiries such as mine and I received no response to a letter I addressed to Nicholas Murray, the Church's Administrator on 12 February 2015. I had requested an opportunity to discuss the illustrations with a representative of the church as follows:[5]

> Further to my letter of 2nd February and the telephone communication from your French colleague yesterday, I understand that the work currently being undertaken by your archivist on the Church's historical resources prevents any study of such material for the time being.
>
> However, I am sending you a copy of my book 'The Hidden Chapter' as a gift to the church. I have indicated the pages which are concerned with Jean Cocteau. Perhaps the archivist will deal with it appropriately.

I had written in an earlier letter:

> I am working on an original ms which I believe is contemporary with the installation in the church of the Cocteau mural in 1959.

My understanding of '1804' at this time left me feeling uneasy, and because I was unable to discuss my purchase with those who might have been able to provide more information I decided to put the project to one side. I knew that I had not arrived at a complete understanding of it in 2015. But after having spent some years researching the 'Stem of Jesse,' it was time to look at it again. It was the role that Tintern played in the saga of that picture which compelled me to study this '1804' masterpiece further. Was it a hoax of some kind? There is a link between the two which had to be addressed in connection with the church of St. Michael's in Tintern.

In Cocteau's church mural I had identified the male figure with the 'fish-like' eye as Sir Percy Loraine (extreme right). This had already been stated in my book "The Hidden Chapter." He was a friend of Cocteau and a close relative of John Lorraine Baldwin, who died in 1896 and was buried in St. Michael's Church. Baldwin was Warden of Tintern Abbey from 1870 until his death. He was also a close friend of the Duke of Beaufort. Both Sir Percy Loraine and John Baldwin were descendants of the prestigious European Lorraine dynasty. The Lorraine family were historically associated with the 'Prieure-de-Sion's' published documents. Again, there are representative drawings of the 'Cross of Lorraine' in the Byrom collection of drawings. The information I have gathered from the collection itself I do not regard as misinformation. I therefore could not ignore what at the time seemed to be a coincidence centred round Sir Percy Loraine. I had some years before identified him in Cocteau's mural. But he is not identifiable in the '1804' illustration. In his place is the figure of Napoleon Bonaparte, crowned Emperor the same year (1804). I conclude that the artist Cocteau chose not to include him in this visual oration because Loraine was still alive at the time. Sir Percy died in 1961. As noted, Bonaparte replaced French 'Royalty' with himself in 1804.

Some years before these discoveries (2011), Michael Baigent, co-author of "The Messianic Legacy," allowed me to reproduce a photograph that he had taken of Cocteau's own tomb mural at 'Milly-la-Floret' in France, which I refer to in "The Hidden Chapter." Jean Cocteau died 11 October 1963, two years after Sir Percy Loraine. I had earlier identified Sir Percy as being the figure reproduced on the triangular mural behind the grave of Cocteau, which Cocteau had personally prepared before his own demise. Sir Percy Loraine's father was himself responsible for researching and publishing privately his family 'Memoirs.' I had managed to acquire access to that detailed document and include here a passage from my book "The Hidden Chapter.":[6]

Chapter 17 – 'Pedigree and Power' p.233
The de Bars later married into the ducal house of Lorraine and in the first chapters of his *Memoirs* Sir Lambton Loraine (sic) goes into some detail about his family's connections with the house of Lorraine and the de Bars. Such genealogical claims would intrigue Cocteau. Since Cocteau was consciously using geometrical forms for symbolic purposes in his art, I felt

I should look again at his relations with Percy Loraine. Did their friendship include a shared interest in the esoteric tradition? Cocteau's use of dreams and the unconscious in his work is well documented and in keeping with an interest in the occult. But it was the age-old symbol of the triangle on the chapel wall at Milly close to his tomb that intrigued me the most – as it is intended to intrigue everyone who sets eyes on it. The triangle's rich, symbolic associations convinced me that Cocteau's prolific imagination and keen intelligence would find fertile soil in the pedigree of a friend like Percy Loraine, whose family claimed descent through the Dukes of Lorraine from Emperor Charlemagne (who married a Merovingian Princess).

I looked again at the head of the figure at the bottom right of the mural in the Church of Notre Dame in London and the head of Christ in the mural above Cocteau's tomb at Milly-la-Floret and noticed a similarity between them. Painted in the same year, could the two heads have been modelled on Percy Loraine? I had been able to study two photographs of Percy. One, taken from the family *Memoirs*, shows him as a young man in army uniform and is reproduced for the reader (illustration 38). The other is a portrait by T. Geraldy painted in Paris in 1913 when, at the age of thirty-three, Percy first met Cocteau. The photograph of this portrait is still in copyright and unfortunately I have been unable to trace the descendants of the executors of Sir Percy's will for permission to reproduce it, but it can be seen in Gordon Waterfield's biography *Professional Diplomat*. If the reader studies these two photographs, particularly that of the portrait, he may be able to see the likeness I thought I had recognised. As the leaflet from the London church reminds us, Cocteau's gift for likeness was 'quite remarkable'. Was this a final tribute from the artist to a man whose abilities he admired and respected?

I would like to have discussed the Milly-la-floret photograph's significance with Michael Baigent, having later acquired the '1804' illustration. Unfortunately, Baigent died 17 June 2013, two weeks before the Rev. Neville Barker Cryer, and a short time before '1804' came into my possession. The Tintern connection persuaded me to pursue the Cocteau line of enquiry regarding the authenticity of this complex piece. As stated in the last chapter, two earlier listed Grandmasters of the 'Prieure-de-Sion' from the fourteenth century, Edouard de Bar, 1307-36 and Jeanne de Bar, 1336-51, were directly associated with St. Michael's Church in Tintern and the Merovingian legacy.

Sir Percy Loraine had a distinguished career as a long-serving diplomat before retiring to breed racehorses. He eventually died aged eighty years without an heir. He was very much respected in the British aristocratic social scene, as well as by the Foreign Office. So how was Cocteau commissioned to paint such a mural at Notre Dame de France in the first place? Publicity information includes some information in connection with this question:

> It was Monsieur Rene Varin, the cultural advisor in London who conceived the idea of asking him to take part in the decoration work of the new church of Notre Dame de France, which was then being rebuilt. Indeed, the first church had been very badly bombed during World War II and the whole of the French community, headed by the ambassador Monsieur Jean Chauvel, was tackling the task. Cocteau's films were enjoying a huge success in London at that time and the artist had to be protected from the invasion of reporters by a high wooden scaffolding all around the chapel!

The same French Church leaflet describes Cocteau's ability:

> Cocteau had a natural gift for drawing. This may have been passed on to him by his father, an amateur painter. Cocteau's dexterity was quite remarkable, as was his gift for likeness. His drawing is very linear. In fact, he compared writing to a kind of drawing and observed: "Writing is a kind of drawing knotted in a different way, and drawing is another use of writing. And when I draw, I'm writing, and perhaps I'm drawing when I write."

It is time to look at the remaining unanswered questions regarding '1804.' Despite its sketchy appearance and in some ways careless use of pen and paper the construction is based on the immaculate geometry of the Pythagorean pentagram. The centre of the inverted C and 'PS' are the focal point of its construction, with angles of 72 degrees positioned from the secretive dot at the top of the canopy. The five points of the pentagram when joined reveal startling elements.

When the points of the pentagram have been joined by straight lines, the lines drawn through the stained part of the page reveals the rarified symbol of the 'Rosicrucians,' appearing directly and purposefully in position on two of the angles. The sign stands for 'as above so below' in Rosicrucian symbology. In Hermetic beliefs, to which the 'Rosicrucians' subscribe, the phrase indicates that earthly matters reflect what happens in the 'heavens' and that members of the Rosicrucian order possess a secret wisdom handed down from ancient times.

Its alternative name is the 'Order of the Rosy (or rose) Cross,' which has close associations with the 'Priory of Sion' (Prieure de Sion). It has been stated that the order:

> Reassumed the mantle by which it had undertaken transactions, both fiscal and temporal during the 11th to 16th centuries, albeit a variation on the name, in keeping with modern Europe.

Prior to 1925, and in the 19th century, the 'Order of Sion' had not made itself public, other than by its spiritual/esoteric body: the 'Ordre Rose and Croix Veritas.' It is the 'trunk' of what is now known as the 'Priory of Sion' and it is the same 'society' which held the 'Salon de la Rose-Croix' in Paris in the late 19th century (headed by Josephin Peladin).

At the bottom right of the illustration is the figure I previously identified as Napoleon, with the famous curls in his hair. The straight line from the "P" in Paris extending to the bottom, intersecting the line going up to the dot at the top making a 72 degree angle, passes through the '70' – the age of Cocteau when the work was being created in 1959. On the lines are symbols associated with aristocratic Estates and families in sympathy with the causes implicit within the illustration, which is the injustice done to the Christian message and the historical persons who tried to present it and have been supressed. It is my opinion that the origin of the Priory and other organisations were devoted to revealing 'truth' to the discerning eye. The 40 below the 20 was the number of years Cocteau had been Grandmaster of the Priory. It will be noted that on the other side of the illustration, to the right, the echoing geometry does not include any such symbols. To the left of the figure of Napoleon, with script nearby, standing on a plinth of seven steps is the Frenchman Bernard-Raymond Fabre-Palaprat. Born in 1773, on 4 November 1804 he founded the 'Order of the Temple' and disclosed the existence of the "Charter of Transmission," which was written in Latin in 1324. It included a list of twenty-two Grandmasters of the Knights Templar, including his own name as the last in 1804. That list differs from the list of Grandmasters accepted by most authorities.

It is significant that the 'Order of the Temple' was founded three weeks before the inauguration of Napoleon as Emperor and that Fabre-Palaprat is positioned on the penultimate (next to last) step of the seven to the left of Napoleon. The son of a surgeon, Palaprat was ordained as a priest but left the priesthood. He then studied medicine at Montpelier, qualifying in 1798 and obtaining a further medical degree in 1803. He claimed to be in possession of personal artefacts in connection with Jacques de Molay, the Grandmaster in 1307 at the time of the arrest of the Templars and the extinguishment of the order. Jacques de Molay was burnt to death in March 1314. Palaprat's inclusion in the '1804' illustration must be of significance.

Palaprat's legacy, in reminding us of Jacques de Molay's immolation in 1314, brought my recollections back to Tintern. King Edward I had died in 1307 as had his daughter Joan of Acres. It was his two grandchildren, brother and sister Edouard de Bar and Jeanne de Bar that later became long-term Grandmasters of the 'Prieure de Sion,' according to the accepted list of the society. As property owners, their association with the Tintern area and St. Michael's Church is indisputable and the burial of John Lorraine Baldwin and Sir Percy Loraine at the same site hundreds of years later cannot be ignored.

The area had some sort of magnetic pulse which through the centuries drew descendants of families and personalities I have studied back to the area. In ages when technology was not sufficiently developed to provide the services of today, traditions, word-of-mouth and heritage were the order of the day. A noticeable number of gifted individuals were able to record matters of history on a single 'piece,' e.g. paper, card, tapestry or canvas, even carvings and murals on a wall. Such was '1804.' The artist here is presenting an accumulation of data not simply from one period of time, but many.

An additional connection with Tintern persuaded me to take '1804' even more seriously. John Loraine Baldwin was the subject this time – the occasion of his funeral. I included the full account from 'The Chepstow Weekly Advertiser' of 5 December 1896 as a quote in "The Hidden Chapter." My reason for doing so at the time can be measured by the following extract from page 85 in my book. It was the individuals in attendance who I felt deserved my special attention, those who had deemed it proper to be in attendance that day:[7]

> A royal prince (George, Duke of Cambridge) and a clutch of coronets were proof enough, if more were needed, of Baldwin's social eminence. Setting aside his grandson of George III (until very recently Commander-in-chief of the British Army), some of those coronets gleamed brighter than others. One in particular caught my eye with a dazzling relevance. The Earl of Cork was Richard Edmund Boyle, the ninth Earl. It was his famous ancestor, Robert Boyle, who had once owned the brass plates from which the Byrom geometric drawings were printed. It was those drawings that had brought me down to Tintern originally in my pursuit of their provenance and the site of the Tintern brassworks. I had discovered that Francis Bacon had been a shareholder and one directly involved in running the wireworks. The way the wheel had come full circle seemed almost incredible. My sense of Bacon's involvement with the 'King of Clubs' and his burial increased with the thought of a Boyle standing by his graveside.

In the published list of Grandmasters of the Prieure de Sion, Robert Boyle is included for 1654-91, a period of thirty-seven years. That list had been taken from the 'Dossiers Secrets.' The compiler seemed to be familiar with personalities throughout the centuries who had associated connections with this beautiful little hamlet. We have noted others from the list earlier. Sir Percy Loraine's direct bloodline connection with Baldwin encouraged me to believe that Cocteau's respect for the Lorraine family heritage was made permanent in the London French church and his own final resting place in France.

And so, on to the hand-written language and symbols section of '1804.' To repeat what Cocteau is known to have said about his own style in representing his art form:

> Writing is a kind of drawing knotted in a different way, and drawing is another use of writing. And when I draw, I'm writing, and perhaps I'm drawing when I write.

When I came to study this section of '1804,' I thought that the style typified Cocteau's thoughts and personal style. Most of his published creative writing is in French, as are his personal diaries. His linguistic skills echo his native tongue. It became evident on close study of '1804' that considering the use of special symbols, my transcription and translation would have to be tentative in places. It would be dishonest to say otherwise. I have studied various codes and shorthands over many years and would suggest that there is no regular or systematic style in use in this piece. But there are certain observations which can be made. Many words and letters on both sides of the illustration have been reproduced back to front as in the inverted 'C' in the central 'PS' figure.

A section of the script (top right) carries information connected to the Knights Templar's list of known Grandmasters, though the spellings do not conform entirely to the registered entries in the records. I refer the reader to the "Knights Templar Encyclopaedia" by Karen Ralls. In her register, dating from 1136-1149 'Robert de Craon' is number two on the list, while number five is 'Andre de Mountbard,' 1154-56. It might appear that the names are similar, comparing that list to the illustration, but the numbers which follow do not represent compatible dates and the names are not exact. I think that the numbers represent catalogue entries to French Government Research papers. The small section with the heading "Le Saint – Graal" translated as 'The Holy Grail' is formed in its own exclusive space, separating the paragraph from Palaprat and Napoleon. Both personalities are associated with the year 1804, as is the execution by guillotine of Cadoudal and his eleven supporters. The '7' in the address "7 Rue de Guillotine" can be connected to the 7 steps, (bottom right). Palaprat is standing on the 6th step, indicating that he is not considered worthy of being on the 7th by the artist. The characters between them are associated with the Fouquet account.

We must now turn to the section on the left. Jean Cocteau took over as Grandmaster of the society in 1918 at the end of World War I. Though this section is dense and complex in its detail, I consider that the lines of script are linked to the decision-making process to do with Cocteau and the organisation itself. Cocteau's legacy of artwork and literature and film reflect an exotic lifestyle of luscious extravagance and a genius to go with it. It is not my intention to comment too much further on Cocteau or any of his legacies. However, it seems that a statement is being made (bottom left) regarding the 'Red Cross' Organisation that had never been mentioned by the artist himself to my knowledge, which he wished to include in this piece. "Never b/e Catalogued – never before catalogued – Croix Rouge". I must include here a note about this phrase.

In 1968, five years after Cocteau's death in 1963, Frederick Brown, the American author, published his masterly biography of Jean Cocteau "An Impersonation of Angels." His perceptions and acquired data illuminate some of the drama of Cocteau's life, recorded afterwards in the diaries of others and subsequently shared by those closest to him after his death.

Such were the circumstances regarding the 'Red Cross.' Cocteau was declared physically unfit for military service for World War I at the age of twenty-five in 1914. Anxious to contribute and sporting an elegant pilot officer's type of blue uniform he became an ambulance driver for a supplementary supporting group to the 'Red Cross' in Paris. On one mission they were to deliver medicine to Rheims, which had been heavily bombed. Rheims had been the coronation city for French Kings since the Merovingian King Clovis I, rather than Paris. The Cathedral was also damaged during the conflict. It was late August 1915 at the time of the incident. Cocteau was said to have acquitted himself well, helping to keep up the spirits of his colleagues, but he was badly injured and suffered a dislocated joint. When he appeared on crutches at a group meeting afterwards he was thought to be a fraud. The army doctor at the time confirmed that his limp was genuine and that his disability was justified.

The supplementary supporting group of ambulance drivers was set up by the wealthy Rumanian aristocrat Princess Bibesco, otherwise known as Mme Edwards, a close friend of Cocteau's during his years in Paris. She kept very detailed diaries of her colourful and eventful life. The Rheims episode made such an impression on Cocteau that he chose to record it in '1804.' The phrase "Croix Rouge" helped to persuade me that '1804' was Cocteau's own work.

Returning to '1804' (bottom left), positioned on the same line as "Croix Rouge" are the words "Pot H" with a symbol between. Oddly positioned, almost out of the picture, I think he is referring to 'Mrs. Henry Pott.' I link this phrase with the line containing the word 'Northants' farther up the same complex portion of the script.

Considering the passage, including 'doodlings,' I think 'Northants' is referring to William Compton, 6th Marquess of Northampton, who was the owner of the historic building in London, Canonbury Towers in 1959, the significance of which is explained below. He was a very distinguished professional soldier during the First World War, receiving amongst other accolades the 'Distinguished Service Order' and then becoming a 'Commander of the Order of St. John of Jerusalem.' His ancestor, William Compton, 4th Marquess of Northampton had been made a 'Knight of the Garter' in 1885 on 9th July. On 18th December that same year, Mrs. Henry Pott started the 'Francis Bacon Society.' Mrs. Henry Pott had devoted her life to researching the life of Francis Bacon and published her authoritative work "Francis Bacon and his Secret Society" in 1891. She is known to have held several meetings through the years at Canonbury Towers. The Rev. Neville Barker Cryer had given me, many years before his own demise, his heavily annotated personal copy of this remarkable study. Jean Cocteau was in my opinion privy to and fully cognisant of the Fr. B. Society and its use as a venue by her and her group. Bacon had leased it for ten years himself during his own lifetime.

I think '1804' reflects the deeply held convictions of the artist. It is also a chronicle of events associated to and reflecting the harsh realities of history. There are other details connected with the seemingly questionable quality of presentation, which I choose not to discuss here, because the understanding is not clear to me. However, some certainties are clear. There is a legacy which is guarded by the Prieure de Sion, a group whose identity remains enigmatic. In stating that truth, Cocteau leaves a clue on the mural of his own grave memorial at Milly. I believe that Cocteau was of the opinion that the Merovingian genealogical lineage preserved truths concerning Christianity that had to be protected.

As a permanent reminder for posterity, he included the physical likeness of Sir Percy Loraine on his own burial-site mural, with Sir Percy depicted as Christ. Cocteau met Percy Loraine at the British Embassy in Paris in 1913 when Sir Percy was thirty-three years old. The number is one of great significance to many traditions, including Christianity. In that depiction, Percy is wearing Jesus' 'crown of thorns' with his 'lineage' arm cupped protectively. Here, Cocteau provides another clue. Positioned strategically on the arm are three groups of three dots, making nine in all. With this device, Cocteau is respecting the 'nine worthies' of Templar legend considered to be the greatest warriors in history. They include Baldwin I, King of Jerusalem. Three of the worthies were selected from the Old Testament, three were Pagans and three were Christian rulers, Godfroi Bouillon, King Arthur and Charlemagne.

There were originally nine Templar knights that formed that group in the twelfth century to protect Christian pilgrims in Jerusalem. Part of their legacy through the centuries is being maintained, albeit exclusively, by the 'Prieure-de-Sion.' Cocteau chose to honour his own memory as a

Grandmaster by recognising and recording the 'House of Lorraine' in his work.

* * *

The Ebay 'bidders' account of the sale of '1804' includes what would be described as Masonic number symbolism in the bids, one consisting of the unlikely amount of £17.62p. This I have interpreted as the year 1762. It was the year the famous Admiral George Anson died. The Anson family were resident in Shugborough Hall. On the grounds of the estate there is a marble bas-relief reproduction of Poussin's 'Les Bergers d'arcadie' in reverse. It had been commissioned by the family around the same date, 1762. The seller was linking the '1804' sale bid with the date of that painting by Poussin with the memorial and what it represents, for the purpose of bringing it to my attention. Why? All I can say is: "Story of my life!"

One must wonder for whom '1804' was prepared in the first place. Where had it been for fifty-four years and who decided that it should be sold on 'Ebay' in 2013? The Bible from which the page was taken would be a step in the right direction. The current whereabouts of that Bible remains part of the ongoing mystery. The only certainty concerning these questions is that the sale was timely and calculated.

I have chosen to include '1804' and my understanding of it at this stage of my research because of my interpretation of the painting 'Stem of Jesse.' Again, there is a connection between the two. The artists of both works possessed knowledge not readily available to the public but which they wished to put on record. As a result of the passage of time and the trafficking of knowledge through modernisation, some records become lost to memory and understanding.

In the next chapter I discuss another piece of artwork. It too was a surprise to me and unexpected; an original sketch in pencil, doubtless thought to be of little real significance at the time. It has added another dimension to my understanding of the previous three 'allegories' and their place in the enigma of the history surrounding Tintern and its relationship to the 'Hermetic' knowledge I have been investigating for so long.

CHAPTER FOURTEEN

Pictures Become Records

B Y THE TIME I was given the pencil sketch, I had studied historical records connected with Tintern and Monmouthshire in some detail. The gift was a generous gesture from a well-meaning supporter who thought to provide more background material which might be of interest. The drawing was of a building which had 'Tintern' written on it, as well as the year '1850.' There was also the signature of the artist, 'L. L. Razé.' Although I knew the area reasonably well I did not recognise the building or the landscape as depicted. On my next visit to Tintern I took the sketch with me and showed it to Elsa and Adrian Wood of The Nurtons. They recognised it immediately and took me to the site.

It was part of the 'Anchor Inn' grounds, a commercial enterprise very close to Tintern Abbey. When the sketch was drawn it was of the 'Ferryman's cottage' and the landing wharf for river traffic to the Abbey and local hamlet. I later learned it was sometimes referred to as the 'Abbey Watergate Slipway.' On old maps it is described as 'The Duke of Beaufort's Boathouse.' The river Wye is no longer regarded as a means for transport and the Ferryman's cottage was by then no longer to be seen. The site had become part of the private grounds of the successful Anchor Inn. It seemed not relevant to the work under discussion here at the time, so another file was created, and the drawing was put to one side.

While reviewing the research data of the three 'allegories' previously presented, I became curious as to who the artist of this pencil sketch of the Beaufort's Boathouse was. The surname 'Razé' was an echo of a name associated with the Prieure de Sion. Razéz had been the name of families associated with Merovee and the Merovingian royal blood line so central to the Priory de Sion saga. So who was the artist, why was he in Tintern in 1850 and what was his interest in the 'Boathouse' and wharf? As my thoughts and inquiries proceeded, what seemed to be a straightforward set of questions produced information which I found tantalizing.

Despite the hastily prepared appearance of the sketch, the artist was a very skilled talent with a genealogy worth looking into. The name had piqued my interest. According to the Henri Lobineau documents published about the Merovingian Dynasty, Giselle de Razés was the aunt of the Duke

Tintern 'Abbey Watergate slipway' L.L. Raze

of Razés who inherited property around Rennes-le-Chateau around 681. The papers state that the Razés family were descendants of Clovis I and Merovee.

There is no 'z' or 's' at the end of the artist Razé's signature; he was a Frenchman by birth. That intrigued me because of the nature and findings regarding of the three 'allegories' I had been studying. Louis Laurent Razé was born in 1804 and came to England at the age of nineteen years with recommendations as an architectural draughtsman and 'teaching abilities.' He became a part-time teacher at the King's School in Canterbury (1842-65). A rare copy of a collection of his 'Views of Canterbury and Europe,' published by the Museum and Galleries of Canterbury, demonstrate the unique talent of this little-known artist. A note at the end of this published collection of his artwork declares: [1]

> Family tradition stated that he was caught up in the Siege of Paris and died shortly after from the effects. His death was recorded on 22nd December 1872 at 141 Boulevard Perire, Paris.

What was Razé doing in Tintern in 1850? I needed to know more. The drawing was given to me in 2012.

Speculation and surmise about the talent and lifestyle of Monsieur Razé could not become a pre-occupation I was prepared to spend too much time on. Already there were matters concerning Tintern that were not satisfactorily concluded, which required my continued attentions. My book "The Hidden Chapter" had been published in 2011, yet my interest was too piqued to ignore it. It was therefore a question of drawing information from official records to fill out the picture.

Fortunately, the archives of France's population are documented in the form of government records with a considerable number of details. In the interest of accuracy, I include the complete entry (translated) of the birth of artist Razé obtained at the time of this research:

Reference:
(cote)(library shelf mark)
5 mi 12 R036 (micro film code) (1st page. complete reference book opened 1800-1804) (Law – 1792 everything has to be registered) n. 573 of the 17th day of month Fructidor the year 12th of the French Republic. Third hour of the opening:

Birth Certificate of Louis Laurent Razé born the 15th day of this month at half past four in the afternoon – son of the Sieur Jean Philippe Razé merchant of dyes and Marie Josephe Jesu, his spouse living Rue de Cantimpretz in Cambrai; the sex of the child was recognised to be male, first witness Francois Delcourte coal labourer (miner) aged of 22 years, living in Cambrai; second witness Pierre Gille, Inn Keeper/landlord aged of 72 years living also in Cambrai; on the requisition to us, done by the Sieur Jean Philippe Razé, father of the child, who signed after reading; the witnesses declared not knowing how to write, established by myself Pierre Joseph Douay, Mayor of the City of Cambrai, acting as public registrar of the undersigned civic office.

Razé Of Douay File
(opened this micro/register 23/Sept. 1800 – to 22 Sept 1804
Left & right
*1091 – microfilm.

Born in the fateful year of Napoleon's coronation '1804,' my curiosity was on the alert. Louis's father Jean Philippe Razé was a 'merchant of dyes.' His mother's name was 'Marie Josephe Jesu' until her marriage. What had persuaded Louis as a talented teenager to settle in Canterbury? He had received a good education and his parents were Christians who had nurtured him in the faith. The family name was to provide an additional clue to my developing interest.

It was the parent's marriage record that provided a possible answer to this shadowy figure's visit to Tintern in 1850. On the marriage certificate Louis's father is described as 'Jean Philippe Razéz.' The extra letter 'z' is clearly significant. The certificate states that he was born at Valenciennes. I include the complete entry, as translated:

Wedding day entry for the parents of : Louis Laurent Razé. No. 16 of the eighth day of the month Frumaire (November 30th 1803) of the year 12th of the French Republic 6.0 clock in the evening.

Marriage certificate of the citizen Jean Philippe Razéz aged of forty one years and ten months, the profession of dyer born in the city of Valenciennes from the department of the north the 24th day of the month of February the year 1762. Living in Cambrai, same department, widower of Marie Josephe Mahout. Son of deceased Jean Francois Razéz, merchant Frippier (clothes) and of arare(?) Therese Disarthe his spouse on one hand. The citizen Marie Josephe Jesu aged of 33 years born in the parish de Sin the district of Donaij of the department of the north the 19th of the month of September the year 1770. profession cook also living in Cambrai, daughter of deceased Jean Antoin Jesu, gardener and of still alive Marguerite Pierron his spouse, living in the so-called parish 'de Sin' on the other end. The preliminary certificate is extracted from the register of the marriage banns done in Cambrai the Sunday 28th Frumaire and the fifth day of this month publicised the present announcement by law without opposition. The birth certificate of the future spouses is accurate and proper, and these acts have been read by me, public registrar. The so-called future couple present, have declared taking marriage, one Marie Josephe Jesu, the other Jean Philippe Razéz in presence of the citizens Charles Thiebaut tailor aged 40 years, Francois Parent, innkeeper aged of 41 years, Daniel David Dyer aged of 45 years and Jean Baptiste Moutiers, merchant Frippier aged of 53 years, all friends of the contractors and living in Cambrai. After what I Pierre Joseph Douaij, mayor of the City of Cambrai acting as Public officer of the Civil State has pronounced that in the name of the law the so called Jean Philippe Razéz and Marie Josephe Jesu are united in marriage of which I have established the present certificate that the named spouses and witnesses have signed, after lecture, except Charles Thiebaut who declared that he couldn't write, done in the City Hall of Cambrai, the day, month cited above.

Razé Marie Josephe Jessus (sic) F Parent
Daniel Monties
Z Douais File.

Translated by Dr. S Choisy.
Original copy in pencil. This was copied from it 28.8.2012

Jean Philippe Razéz, born 1762 in Valenciennes, must have been sensitive to his family heritage at the time of the birth of his son Louis in 1804, sufficiently so to drop the final 'z' on the formal document registering his son's birth. Numerous historical records going back to the 9th century record the 'Count of Razéz' and their descendants living in the area. They are listed with the letter 'z' as for Razéz. Political activities around 1804 could have been an influence on the father's decision to drop the 'z.' When one observes the detail considered necessary for the records, the registered names of the individuals would be a permanent entry. So Razé it was.

When Louis Laurent Razé arrived in Canterbury in 1825, as may be surmised by his background and upbringing as an artist interested in architecture, he was already skilled in the use of sacred geometry, the underlying principal of Gothic architecture. His accurate representation of Canterbury Cathedral, included in the publication of his 'Views of Canterbury,' is thought to be one of the finest ever produced. Razé died in 1872. The publication is dated 1990, over one hundred years after his death. The late Brian Stewart, responsible for editing the publication, includes the following observation which may account for Razé's somewhat reclusive lifestyle:[2]

Little is known about Razé's life. He was listed as a recusant – a Roman catholic who refused to attend services in the Church of England. He had married by licence on the 10th May 1828, Ann Nisbett of St. Alphage Canterbury [and possibly the daughter of Henry King Nisbett, a leather cutter]. The couple had two daughters, both artists, Emilie and Angelique.

Having been brought up in Cambrai, Louis would have known of the destruction of the City's beautiful Gothic cathedral in 1796. Already damaged, it was converted into a granary. The spire was blown down in a storm in 1809. Louis would have been five years old then. This Cathedral had been consecrated in 1472 and was known as 'the wonder of the low countries.' With its classic Gothic measurements of 131 metres in length and 72 metres wide, the singing within was the 'finest,' the bells the 'sweetest.'

With the parental support and traditions associated with his family background, I think it very likely that Razé came under the influence of Charles Nodier, the French Grandmaster of the Prieure de Sion when he was a teenager. Known for his love of Gothic architecture, Nodier became the librarian of the 'Biblioteque de l'Arsenal' in Paris, a position he held for the duration of his life. The library was founded by King Francois I in the sixteenth century, now part of the National Library of France. Nodier was a lover of literature including the works of Shakespeare. He became Grandmaster of the Prieure de Sion in 1801 and remained in that position until his death in 1844. What I found particularly interesting was his known association with a secret society described as 'Biblical and Pythagorean.' Louis Razé's years in England must have been culturally bound up by principles cloaked in secrecy, otherwise much more would be known of his life and his time in England.

When I recognised the exclusive talent of Razé the artist, represented in the collection of his work in the Canterbury booklet, his visit to Tintern became even more intriguing. He arrived in England in 1825 and died in 1872. From census records it became clear that there was a span of at least thirty-five years in the United Kingdom to be accounted for.

Listed as a 'recusant' who refused to attend services in the Church of England, one wonders how he obtained a teaching position in the heartland of the Anglican church, Canterbury, when non-conformists were barred from so many offices. When he visited Tintern he had been in the country for twenty-five years. It seems that Razé became a resident at Canterbury in 1825 at the age of nineteen. Three years later he married a local girl, Ann Nisbett on 10 May. They settled in the community despite his status as a 'Recusant.'

The Archbishop of Canterbury was the grandson of the 3rd Duke of Rutland, a direct descendant in the line. The 3rd Earl had acted I believe as a 'double' for Francis Bacon as previously mentioned. The same family owned Fisher's Folly in London, which was the venue of that fateful dinner for the 3rd Earl of Pembroke on 9 April 1630. The Pembrokes were Patrons of St. Michael's Church, Tintern and the Archbishop at the time of Razé's arrival was Charles Manners-Sutton. Manners-Sutton would have known of the historical background of his own family's heritage. His grandfather was succeeded by the 4th Duke of Rutland, who was married to Lady Mary Isabella Somerset, sister of Henry, 5th Duke of Beaufort, the custodian of Tintern Abbey. It is my opinion that these dynastic links were very much a part of Razé's reason for being in the neighbourhood of Canterbury and Tintern and for the mysterious silence surrounding his activities.

I realized it had to be public records which would provide information to fill the gaps in the years of Razé's time in Canterbury. The census for 1851, the year closest to his visit to Tintern, states that he and his wife were resident in Canterbury with one servant. There is no mention of his two daughters, Emilie born 1832 and Angelique born 1834. A later census for 1861 clarified the family status of the two daughters. They were both living with their parents, Emilie aged twenty-nine years, her sister Angelique twenty-seven years. The address is given as 'No. 3 Victoria Terrace, Canterbury.' Louis is described as a 'Drawing Master' and a 'naturalised British Subject.' By 1861 they had two resident servants.

Both daughters had been described as artists in the booklet of their father's work and in 1861, although both were in their latter twenties, neither was married. But in 1862 the situation had changed for the younger daughter, Angelique. She married in Paris a Frenchman by the name of Jules Andre Durand. The records show that they lived in the Asnieres-Sur-Seine area, approximately six miles from Saint-Denis, part of the northern suburbs of Paris.

The public records show that Angelique and Jules Durand brought up a family of four children in that same area, two sons, Andre Jules and Arsene Eugene and two daughters Gabrielle Victoria and Angelique Gabrielle. They in turn married and raised children in the Asnieres district. Their stability and settled lives indicate a unified family loyalty of long standing. The names given to the younger generation, together with the occupational livelihoods of the adults provide further considerations to be pondered.

The Durand family names of the period, such as Bernard, Denis, Clovis, Mary Magdalene, Rene, Germain, as they were recorded implied a respect for the Basilica of Saint-Denis, which as we know was the final resting place of the Merovingian monarchs. The last king to be buried in Saint-Denis was Louis XVIII. During the French Revolution 1793-1803 the royal necropolis was looted and destroyed, including the tombs. The contents were reburied in a 'common assuary,' to be returned to the abbey in 1819.

Louis Razé's parents were married in 1803. The Razé family with its Merovingian heritage would have been very much affected by the political influences and changes in France at this time. By 1860 Paris had become enlarged and amongst its additional communes was 'La Chapelle-Saint-

Denis.' The working occupations of Angelique Durand's family and immediate descendants are recorded as a carpenter, coachman, tapestry makers, a polisher and a printer. All lived in nearby Asnieres-Sur-Seine. Could they have been part of the maintenance staff of Saint-Denis?

I may not have continued to look so closely at the French records if it had not been for the absence of any record of Louis Razé or his wife in the 1871 census in England. He died in December 1872 in Paris as a result of being 'caught up in the siege of Paris.' The siege took place between 19 September 1870 and 28 January 1871. The political implications were complex. There were thousands of casualties. The result of the siege was the 'Proclamation' of the German Empire. I have not been able to locate evidence of the role Louis Razé played in the campaign, but he must have relocated to Paris sometime between 1861 to 1872. He died at 44 Boulevard Perire, Paris.

Although I was of the opinion that Razé's visit to Tintern in 1850 was connected to the Merovingian history of the area and St. Michael's Church, factual evidence again was scarce. It was the district of Asnieres-Sur-Seine which provided a plausible reason. It was the area chosen for or by his daughter Angelique and family to live. Situated in the vicinity was the stately home of 'Chateau d'Asnieres,' one of the gracious estates near the centre of Paris. The district had been lived in by some of the most talented and influential artists and bankers before the Razé and Durand families settled there. One of the residents had been the beautiful Anne Marie de Bourbon, a daughter of the Prince of Conde and the Bavarian princess Anne Henriette of Bavaria (Enghien) of the reigning House of Bourbon. The official title of her parents was the Duke and Duchess of Enghien. Sadly, Anne Marie died of lung disease, unmarried in the year 1700. But her beauty, not only of appearance, but personality, kindness and generosity left a lasting impression on all who knew her in the district in which she lived. The House of Bourbon was certainly a centre of turmoil of conflicting loyalties for the next hundred years and more. Louis Razé became a part of that during his lifetime and suffered as a result.

A direct descendant of Anne Marie de Bourbon, another Duke of Enghien, Louis Antoine became another victim, this time centred round Napoleon I at the beginning of the 1800's. The young Duke, born 1772, had served in the Conde Armies of his father and grandfather, distinguishing himself by his bravery on several occasions. But early in 1804, Napoleon heard that the duke was involved with Georges Cadoudal in plotting against him, wishing to restore the Bourbon monarchy. The Duke was arrested and shot in the moat of a castle on 21 March 1804, aged thirty-one. His last words were "I must die then at the hands of Frenchmen." Napoleon became Emperor 2 December 1804.

The execution caused shock and dismay across Europe. The Bourbon 'Restoration' began in 1815, following the first fall of Napoleon in 1814. When I read of the young Duke of Eughieu's alleged association with Cadoudal and his execution during the fateful year of '1804,' Jean Cocteau's '1804' took on an additional dimension. The network of the Prieure de Sion, by association, held confidences and loyalties which remained silent and supposedly lost, time having intervened if nothing else.

Louis Razé's loyalties to his personal heritage over fifty years later bear witness to that. I have long felt after many years of study that Tintern became a protective haven for data of a special kind, which then became locked behind the doors of secrecy, maybe unwittingly, maybe not. What had attracted Razé to Tintern in 1850? It seems that he had a close association with the Dukes of Beaufort and their families. The 6th Duke of Beaufort inherited those Estates in 1803. His eldest son Henry Somerset, the future 7th Duke of Beaufort married the sister of his deceased wife in 1822. Her name was Emily Smith. The marriage was considered illegal for many years because his wives were half-sisters and related to the Duke of Wellington. The Somersets (Beauforts) lived in Paris for some time. Their eldest daughter was named after her mother, Emily.

Louis Razé came to Canterbury shortly after that marriage in 1824. He and his wife Ann named their eldest daughter Emilie in 1832. Lady Mary Isabella, the matriarchal widow of the 4th Duke of Rutland was still a very powerful influence within the Rutland family during the whole of this time. She died in 1833. We must remind ourselves that she had, as the sister of the 5th Duke of Beaufort, family connections tied to the Archbishop of Canterbury. Razé's allegiance to the ancient Razéz dynasty, the Merovingians, Saint-Denis's (Paris) and then historic Tintern would be intriguing to the Beauforts at this time.

Louis, in naming his daughter 'Emilie,' suggests that he was connecting links made earlier in Paris. However, it is also a fact that within the genealogies of the Razéz family there was a distinguished lady by the name of 'Amelie de Razés, Countess of Aubnay,' said to have been born about 1000 and lived until 1072. She was the daughter of Viscount Arnaud II de Razés and Lucie de Provence. Louis and Ann were apparently preserving connections of one sort or another in the naming of their eldest daughter. It mattered to them.

During this period in Tintern's history, the Somerset/Beaufort family was very powerful in the hierarchy of the Freemasonry Lodge System. The brother of Lady Mary Isabella, 4th Duchess of Rutland, was the most powerful representative in Britain. Henry Somerset, 5th Duke of Beaufort, was elected Grandmaster to the Premier Grand Lodge. He remained in that position for several years until he retired because of a disagreement in policy decisions within the organisation.

The results of my commissioned ground penetrating radar and archaeology projects at the Church of St. Michael's and its graveyard confirmed an ethos associated with Freemasonry and early Templar movements. Several symbols associated with those organisations had been identified on the site. I had to consider whether Louis Razé's visit in 1850 was in some way connected to well-kept secrets in the area. Four years earlier in 1846 St. Michael's Church had undergone major building alterations which included an extension to the main structure of the church.

When Louis made his 1850 visit it was Henry, 7th Duke of Beaufort married to Emily, who was the resident at Badminton, the Beaufort Estate. It was during their earlier years in Paris that Henry, the 8th Duke was born in 1824. The pencil drawing of 'The Duke of Beaufort's Boathouse' could have been a poignant memory for Louis, for it was in 1824 he arrived in Canterbury and stayed, later calling his own daughter Emilie. The 1846 changes at St. Michael's Church may have been a reason for the visit.

The incumbent at St. Michael's in 1850 was the Rev. John Mais, also the appointed Chaplain of Bristol Royal Infirmary. His duties at the Tintern church could only take place three months out of the year because of his responsibilities at the Bristol hospital. There were already associations with the nearby estate of The Nurtons in connection with John Edmonds Stock, the owner, also a doctor at the Bristol hospital. At that time, actually living at The Nurtons in 1850 were Catherine and Thomas Tireman. Thomas was by profession a 'Clerk.' As a Reverend in residence near to the church, he is recorded as performing many of the services at St. Michael's in 1850. The original church records show that he also officiated at the burials that took place in 1846 except one. Given that major alterations were taking place on-site that year, it would have been a sensitive time to be responsible for ongoing work requiring many decisions.

Based on my accumulated research over the years, the Reverend Thomas Tireman and his wife Catherine can be identified as descendants of John, 1st Duke of Athol. That family can be traced directly back to John, 1st Marquis of Hamilton and then King James II of Scotland. Reverend Tireman died in 1852, two years after Razé's visit, and was buried under an ancient Yew tree, situated close to the outer boundary of St. Michael's churchyard, some distance from the recent alterations. The same burial site had been used earlier in 1827 and 1839 for the final resting place of earlier residents at The Nurtons. George Place is described as a 'Gentleman' in the burial record for 20 January 1827, aged 59 years. His widow, Dorothy is described in the same records for 15 March 1839 as of "'Nurton' aged 73 years." The very next burial entry, 18 July, is for Sarah Lewis. The officiating cleric is the Reverend Thomas Tireman. The Tiremans must have moved into The Nurtons very soon after Dorothy Place's burial. Checking the genealogies, George and Dorothy Place were also part of John, 1st Duke of Athol's family's Herbert connection. All seem to have had a specific interest in the area. A Yorkshire man, the Reverend Tireman can be seen in the Church Records as continuing to perform most of St. Michael's church services for the next years with occasional interruptions. It seems that those were times when he was out of the district.

I was fortunate enough to be given access to the original church records for close study. The Reverend Phil Rees, the incumbent at the time, allowed me to do this. I copied them out in their entirety for the years that I was particularly interested in and so was able to note the additional details provided in them by the recording 'clerks.' Rev. Tireman was resident at The Nurtons from 1839 until his own demise in 1854. But in the first National Census taken in 1841, the recorded resident was 'David Evans, 40, Clerk,' with three female servants. Interestingly, The Nurtons is described as 'Nurton Priory.' On 2 May of the same year the resident, David Evans is described as the officiating minister of the burial service of one Emma Brown, aged one year and nine months, at St. Michael's Church. It is the only service that the Reverend Tireman did not perform that year. In no other recorded census since is The Nurtons described as 'Nurton Priory.' The Reverend Tireman had been temporarily absent from the district in 1841, yet he had returned by 16 May. That fact is clear because he officiated at a funeral on that date.

Fortunately, I store my accumulated records and research material from my previous published books and recognised that there were still tantalising questions needing to be tested and answered in a more satisfying way. I have been led to believe over the years, by those older and wiser, that the answers are 'out there if you can find them.' Tintern, for me, had become more than an enigma. I was certain that all I had discovered thus far and presented in published works was part of a larger issue still waiting in the wings to be 'listened to.' It seemed to me as though the soul of the area 'rustled' through the trees, as they stood majestically cascading down the hillsides and through the country lanes. Buildings, some old, some ancient could say so much to an audience if asked.

I felt that Louis Razé was on a quest looking for answers when he made his visit in 1850. Everything I had learned about him thus far indicated that he was something of a unique talent and an extraordinary human being. The detail in his published drawings of architectural buildings suggests a tidy mind. I imagined him to be a rather lonely intellectual. I had noted that there was no representative illustration of Tintern Abbey in his published work. The pencil drawing, I had been led to understand, became part of a 'job-lot' around 1900 in an auction in Lincolnshire. He became for a time something of a 'catalyst' in my approach to his possible ambitions and those ambitions became mine, imaginary or not regarding 1850.

Given Razé's family connections, he would certainly have visited the Reverend Tireman at The Nurtons and St. Michael's Church. The historic genealogical links regarding Tireman's church responsibilities, together with all the factual data I had gathered, suggested that those resident at The Nurtons through the centuries were guardians of heritage, tradition and knowledge of a secretive nature. They were acting exclusively for a group of elected members, I suspect. Aspects of the history of Tintern Abbey and St. Michael's Church were part of that responsibility. The exact purpose of Razé's visit in 1850 remains something of a mystery, but I believe it would have some connection with St. Denis's in Paris and the 7th Duke and Duchess of Beaufort.

I then chose to consider again the inclusion of the Abbot Suger's 'Eagle vase,' which had been part of the religious décor at Saint-Denis's for many centuries before being moved to the Galorie of Apollon at the Louvre in France. The artist Lodge of the Tintern allegory I call 'Stem of Jesse,' had included the vase as an altar piece in the St. Michael's Church section of the painting. Its position was deliberate and purposeful. It provides a clue which joins the two religious establishments together, albeit in different countries.

I decided that the alterations at St. Michael's were in some way part of a question still unanswered. The ancient graveyard had been disturbed. I studied again the results of my commissioned investigations of G.P.R. and the archaeological work done on the site, together with the records of burials over the course of a hundred years in the oldest part of the graveyard. Since acquiring and reporting on the findings initially, my research had been devoted to other challenges. Coming back for a fresh look at the entries provided me with an additional purpose and perspective. Imagine my interest upon learning that during the hundred years of burials between 1813 and 1904 there were only two entries for burials dated '9 April,' the date which had already registered the deaths and other major events in the lives of some of the main figures covered in previous chapters.

Those two entries were to provide another startling development in the Razé story. One of the two '9 April' entries was for the year 1850, the year of Razé's visit. The interment was of a young man of twenty-five years, James Sadler (entry no.1740). I think Razé may have been in Tintern for a commemorative reason, apart for any other purpose. Francis Bacon's officially recorded death date is 9 April. It is also the date in 1630 of the 50th Birthday Celebratory dinner at Fisher's Folly for William Herbert, 3rd Earl of Pembroke. He was Patron of St. Michael's church at the time and the dedicatee of Shakespeare's First Folio (1623). So both Bacon and Pembroke are directly associated with the church site as we know, for that day of the year.

I would not be so certain about the 1850 funeral entry if it were not for the second entry, recording the '9 April' of another funeral. This was for the earlier year of 1843. The entry in the records is 'No. 129.' It records the funeral of a child, 'Hannah Brown' aged four years. The service was conducted by W.F. Audland, the only church service he is recorded as rendering in the whole volume of recorded church services. The very next entry 'No. 130' had already aroused my interest and had been commented on in some detail in "The Hidden Chapter." It was entered for a local resident:

> Ann Watkins. Tintern. May 39th. 33 yrs. G.T. Hall
> Officiating Minister

There is no such date as 'May 39th', but to the Church administration it could be an indicator that this entry was a clue supplying direction. The "39 Articles" is a special document of guidance to all members of the Anglican Church. It will be remembered that Ann Bacon, Francis Bacon's surrogate mother, translated the original from Latin to English. I believe that she is buried in the South Porch of St. Michael's Church. The recorded entry of the death of Ann Watkins states that she was '33 yrs.' at the time of death. 'Thirty-three' also sends out special resonances to Christians and Freemasons. No one had corrected this entry. The site of her burial is also of significance. It is positioned near to the boundary stone wall separating the graveyard from the landing wharf of the River Wye. Within the stone wall there had been set a keystone, another ancient symbolic clue telling knowledgeable witnesses that documents and artefacts were/ could be close by. Ann Watkins' grave was conveniently positioned. These details were noted in "The Hidden Chapter." At the time when I was researching the keystone, I had not included the previous death entry into my considerations. The accumulation of events associated with '9 April' had become a 'constant' in following the lives of Francis Bacon and those immediately associated with him over my years of study.

The entry of '9 April' positioned immediately before Ann Watkins was curious, especially since the officiating clergyman carried out no other ceremony at the church over the following years. Entries No. 129 and No. 130 were considered together. The next recorded burial was for 4 June 1843.

In the 1841 census Audland is described as a member of the staff and resident at Queens College, Oxford University. Quote:

William Fisher Audland. 35. Fellow and Dean.

As Dean, why would he be carrying out the funeral service of a little girl in Tintern on 9 April 1843? There was no suggestion that she was in any way related to him. Who was this man? Again, it seemed to be connected to the commemorative date '9 April.' Audland was listed in published documents for the University of Oxford as a 'Public Preacher for 1839 and 1840, an M.A. Fellow of Queens.' 1839 was the year that the Reverend Tireman moved into The Nurtons. Audland must have been known to the Tireman family because later in October of 1849 he was joining the Reverend Tireman's wife Catherine on a visit to her relatives in Yorkshire. The published diaries of Catherine's niece, 'Jenny,' edited by a later relative Christopher Audland and made available publicly in 2008, was an account which provided yet another insight into that period of Tintern history. I include a quote from Christopher Audland as it appears in his account of Jenny's diary from 1849, Jenny:[3]

> But there was also a significant contact with the family of Mama. In mid-October her sister, Catherine Tireman came to stay at the Rectory for a week.

Christopher continues:

> The Rev. Tireman was then retired and living at Tintern. She [the sister] brought with her a 'Mr. Audland'. The context makes it clear that this was not Jenny's future husband (a General Practitioner at Tintern), but rather his older brother the Reverend William Fisher Audland DD (who was the Rector of Enham in Hampshire, and at that time – or perhaps rather later – a Fellow of Queen's College, Oxford, where Gordon and Eddy now were).

The visit to Yorkshire was a few months before Razé's visit to Tintern. A few months later, on 6 November 1851, the Reverend William Fisher Audland performed the marriage ceremony of his younger brother John to Jenny, at St. Mary's Church in Scarborough. Brother John it seems was already a General Practitioner in Tintern and the surrounding area. Jenny moved to Tintern after her marriage to Dr. John Audland late in 1851. His medical appointment to Tintern was already established. I can find no record of when that was. Neither can I find any reference as to his whereabouts in the census for 1851 in England or Wales. But I am grateful to the Editor of John's newly wedded wife's (Jenny's) diaries, because he included the following reference as a note for the doctor, which I include here. John Audland was born in 1813:[4]

A green trunk (Archives Box 3), located at the Old House, Ackenthwaite in the SE attic under the eaves, contains two scrolls (unwrapped). These are: (a) Certificate of Guy's Hospital, London, that 'Mr John Audland hath attended the Practice of Surgery as a Pupil in this and St. Thomas's Hospital for twelve months last past': signed and dated 14 January 1835; and (b) Certificate of the Court of Examiners of 'the Society of the Art and Mystery of Apothecaries of the city of London...that John Audland has been by us carefully and deliberately examined as to his skill & abilities in the Science and Practice of Medicine ...'; signed and dated 27th August 1835.

'Ackenthwaite,' near Kendal had been the family home of the Audlands for generations.

As I collated material for the period associated with that of Razé's visit, it became even more singular. But then so did Audland's service of 9 April 1843 and that of 9 April 1850. There had to be a link between the two events. The reverend Thomas Tireman was one, but since he was not the official resident incumbent of St. Michael's it had to be a higher church authority. I had already drawn a connection to Charles Manners-Sutton, the Archbishop of Canterbury (grandson of the 3rd Duke of Rutland), who was in residence when Razé arrived in Canterbury in 1824. He died in 1828. He was associated with the Dukes of Beaufort by marriage. Who was his successor in 1828?

The most Reverend and Right Honourable William Howley became the Archbishop of Canterbury in 1828, remaining in the position until his death in 1848. He was an active Freemason, joining the 'Royal York Lodge' in Bristol in 1791 aged twenty-five years and served for some time as Master of the Lodge. A colourful personality, as Archbishop it was his duty to inform Princess Victoria of her succession to the throne in 1838 and earlier in 1831 he had presided over the coronation of William IV and Queen Adelaide. One of his great passions was architecture and he is credited with the rebuilding of Lambeth Palace in the 'Gothic Revival style.' He would certainly have known Louis Razé. Razé's knowledge of the geometry of Gothic architecture is apparent in his published booklet and the fact that he was appointed to 'The King's School, Canterbury' in 1842 bears that out.

Founded in 597 AD as a medieval Cathedral school by Augustine of Canterbury, founder of the English Church and considered to be "Apostle to the English," the school was named King's School in deference to Henry VIII after the dissolution of the monasteries in 1541. Razé remained on the staff of that institution for twenty-three years. His skill and talent would have been appreciated by the Archbishop. He would also have known the Rev. William Audland. According to the Jenny diaries, we must remind ourselves that Audland was the Rector of the church in Enham in

Hampshire. This was but a few miles from Ropley, where Howley himself was born in 1766 and where his father was the local Vicar. Ropley and Enham are both villages in rural England where tradition and families held allegiances for centuries. The traditions associated with the duties of the Archbishop of Canterbury and of the Crown act as a beacon of loyalty to conventions through the centuries.

Although these individuals originated from different areas of Britain and Razé was from Cambrai in France there was some sort of magnet centred round the church that drew them to Tintern Parva in the mid nineteenth century. The Archbishop's connections with Freemasonry in Bristol would inevitably link Reverend Audland with the Dukes of Beaufort, the Somerset family and The Nurtons. I sensed that Razé's 1850 visit had something to do with the major alterations at St. Michael's in 1846. Archbishop Howley died in 1848. So, who was to take his place?

The most Reverend and Right Honourable John Bird Sumner became Archbishop in 1848. Born in Kenilworth in 1780 his birthplace certainly equipped him to understand the historical importance of Tintern Parva and St. Michael's Church. Kenilworth Castle in central England, one of the greatest historical sites in the United Kingdom had been granted in 1563 by Elizabeth I to 'her favourite Robert Dudley, Earl of Leicester.' Dudley then converted the castle 'into a lavish palace fit to entertain his queen.' Elizabeth and Robert Dudley, as reported by earlier researchers, were suspected of being the biological parents of Francis Bacon as well as the Earl of Essex, Robert Devereux. Archbishop Sumner could not have been unaware of the alleged relationship between Queen Elizabeth and her acknowledged favourite, whom she called "Sweet Robin."

The cameos of visual information depicted in these three 'allegories,' when considered collectively alongside other data, implied that there might have been an upheaval of another kind carried out secretly at the time of the extension to St. Michael's in 1846. Why do I sense that? Archaeological records demonstrate that St. Michael's Church had foundations going back to the Roman period, with possible cavities that could have been used for storage. This is apart from its use as a burial site for local inhabitants.

I continued to gather data to do with the church from official records and The Nurtons in particular. The absence of Reverend Tireman at times from the area and the missing local Doctor John Audland from any census record for 1851 occupied a gap which I had no answer to. But it was, as so often before, Sir Thomas Herbert of Tintern, the owner of The Nurtons from 1640 to 1681, who was to provide a missing strand, another link in the mystery of Razé. This time it would be circumstances and chance: Herbert had been the constant companion of King Charles I for several months when he was confined before execution. Herbert accompanied the

King to the scaffold on the day of Charles's execution in 1649. He was also responsible for the funeral arrangements. King Charles had given Herbert some of his most personal possessions, including his own copy of the second edition of Shakespeare's Folio of plays. Herbert had become the 'custodian' of confidences of King Charles, some of which, when placed with others, have been expressed visually in art form and in exclusive manuscripts.

In this instance it was King Charles's wife, Queen Henrietta Maria of France who was to occupy centre stage. Born in the Palais du Louvre, Paris in 1609 she was the daughter of Henry IV of France and Marie de Medici. They too were of the aristocratic house of Bourbon. She was not a popular Queen, being very opinionated, extravagant and a Roman Catholic. Henrietta married Charles at Notre Dame in Paris by proxy on 1 May 1625, shortly after he became King. She would have been sixteen years old. Some weeks after that on 13 June a marriage ceremony was performed for them in Canterbury Cathedral and they spent their first night together at the Royal Palace of St. Augustine's in Canterbury. This was Razé's territory. He taught at the Cathedral School for twenty-three years two hundred years later. He would have known Henrietta was a member of the House of Bourbon and a Roman Catholic.

After the execution of Charles I, Henrietta lived for several years in a convent that she founded in Chaillot in France. She returned to England following the Restoration of her eldest son King Charles II in 1660. She died on 10 September 1669 at the Chateau de Colombes, near Paris, aged fifty-nine years, having taken an overdose of opiates as a painkiller. She suffered badly from bronchitis. Queen Henrietta was buried in the French royal necropolis at the Basilica of St. Denis. Her heart was placed in a silver casket and buried at the convent she founded in Chaillot. Her final resting place was with the early Merovingian Priest Kings and those that followed. Family allegiances remained strong and she was included. Where does this lead?

It was a chance purchase of mine which provided me with what I have come to regard as one of the final pieces of the Razé mystery. It was a copy of a book that had been regarded as surplus to requirements by Tunbridge-Wells library [5]. The title of the book is "King Charles, Prince Rupert and the Civil War." It is based on a collection of letters found under tiles in an old oak chest that had been put up for sale. The owners of the furniture lived in Leeds Castle, Kent. The year was 1822. The oak chest was bought by a shoemaker, who discovered the correspondence under the tiles, consigning it to his own cellar as waste. The letters were related to the English Civil War and though the shoemaker had cut some of the paper into strips, a local banker with an interest in the history of the Civil War purchased the remaining collection of letters. The collection included some original correspondence of King Charles I and King Charles II and Prince Rupert, the son of Elizabeth of the Palatinate, Charles I sister. Eventually the collection was purchased by a publisher, Mr. Bentley in 1847. Later, after various literary exchanges, Sir Charles Petrie undertook the task of editing the collection which was published in book form by Routledge and Keegan Paul, London in 1974.

I bought a much-used copy of this book which takes its place as a treasured possession amongst many others in my library. The index at the back of the book pointed me in the direction of a single entry. It was under the heading 'Stratford-on-Avon.' It was in connection with the year 1643. Queen Henrietta, King Charles's wife, would from time to time travel to venues where the king was conducting his political business. King Charles, always concerned for her safety, instructed his nephew Prince Rupert to keep a watchful eye on her security.

The entry on P. 42 quotes a letter to Prince Rupert, signed 'Falkland.' The letter informs the prince that 'Her Majesty will be this night at King's Norton in Worcestershire ...' the letter is dated '10 July.' The page entry continues:

> At Stratford-on-Avon the Queen met Rupert and was entertained for the night by Shakespeare's now widowed daughter, Suzanna. From there she moved on to the valley of Keynton, near the foot of Edghill, where Charles was waiting for her, and together they passed to Oxford.

Suzanna was to die on 11 July 1649 aged sixty-six years, six years to the day after the visit of Queen Henrietta, King Charles I's future widow. William Shakespeare had bequeathed most of his properties and wealth to Suzanna, who had married a local doctor, John Hall in 1608. Hall was the executor of Shakespeare's will and died in 1635.

Queen Henrietta was known to have favoured the Arts and often performed in 'Masques' at Court. How much would she have known about Shakespeare's personal life? He had been dead for ten years when she married King Charles and came to England. One wonders how the pregnant Queen came to be the guest of Shakespeare's daughter at the time. Suzanna's legacies would have included knowledge and some understanding of Shakespeare's role as an actor and author. But although her signature has been identified on documents, it is not clear whether she received any formal education or just how interested she was in her father's work and legacies.

King Charles I was a confidant of Francis Bacon. Certainly, Thomas Herbert would have known of the Bacon/Shakespeare authorship arrangement. Herbert had been given King Charles's personal copy of the

Folio of plays as a bequest. Queen Henrietta would have understood the truth too. Her visit to Suzanna may well have been for some clarification of a literary matter. It was not simply a light social visit to Shakespeare's daughter, an elderly widow of sixty, or a need of accommodation for the night.

By 1643, the year of Queen Henrietta's visit, Thomas Herbert was the owner of the Hermetic College, based at The Nurtons. He had married in 1632, Lucy, the daughter of Sir Walter Alexander, a Gentleman Usher to King Charles. It was also a fact that at this time John Dee's eldest son Arthur was physician to King Charles. The King by now had become the Patron of St. Michael's Church. It is my opinion that Francis Bacon died in 1632, having returned from Nuremburg after the death of the 3rd Earl of Pembroke in 1630. Records show that St. Michael's Church received serious attention in 1632 with the appointment of William Pritchard, an Oxford Scholar, as permanent Rector. It may well be that the subterranean chambers beneath the church itself and the graveyard became the repository for precious literary archives of Francis Bacon and his closest allies, together with artefacts of historical legacies from previous centuries. There seemed to be something of an exclusivity about the site.

I found Queen Henrietta's visit in 1643 to Suzanna in Stratford as tantalising as Razé's visit to Tintern in 1850. Both were unique in their own way, with a sense of purpose. Two hundred years apart – there was a connection.

St. Michael's Church continued to go through its own period of change after Razé's visit and the death of Reverend Tireman in 1852. In 1870 the Reverend John Mais, the appointed Cleric at St. Michael's Church, retired. The chronicle of events during the next few years record the arrival in 1873 of John Loraine Baldwin as Warden and custodian of Tintern Abbey. He had been appointed by the 8th Duke of Beaufort. A confirmed bachelor, aged sixty-four, he had married earlier that year Lady Frances Russell, a widow aged fifty-one. They were to live close to the Abbey in an attractive residence, 'St. Ann's, Chapel Hill.' Their colourful lives both before their arrival in Tintern and while living there has been recorded in some detail in "The Hidden Chapter." Both were members of the aristocracy.

For the purposes under discussion here, it is necessary to note that Baldwin was a member of the historic Lorraine dynasty with its claim to the family's descent from Lothar the eldest son of the Holy Roman Emperor Charlemagne, crowned in 800. It included such names as 'Godfrey de Bouillon' and the 'de Bar' family. As we have seen earlier, Edward I's daughter, Eleanor, married Henri, Count de Bar whose son Edward de Bar and daughter Jeanne de Bar became Grandmasters of the 'Prieure de Sion' for a total of forty-four years. Charlemagne was married to a Merovingian Princess, which links him to the 'royal' Merovingian bloodline, which links all these characters to Tintern and the Prieure de Sion, under whatever name it might have been called. John Loraine Baldwin died in 1896 and his wife eight years before. Both recorded burials are in the graveyard of St. Michael's, Tintern Parva. Since they lived close to the Abbey and next to St. Mary's Church, one might wonder why they did not choose that nearby church for their final resting place. Clearly, St. Michael's was chosen because of its Merovingian associations.

Another fact must be stated. At the time when my commissioned exploratory G.P.R. investigations were being carried out in the graveyard of St. Michaels by Stratascan Ltd. in 2006 and 2007, the railed enclosure containing the graves of Baldwin and his wife Lady Francis Russell were included in the survey. Peter Barker, the senior executive of Stratascan, was so concerned by the company's initial findings that professional experts in the design and fabric of coffins were consulted. The reason for Barker's concern was that the results of their findings did not match the expected results. Consequently, further scientific investigations were carried out by them at their own behest.

The conclusion was that in all probability the coffins of Baldwin and Lady Russell were not in the location originally placed. The evidence did not match up to the recorded accounts. It seems that the site may have been reused later in a different way. Some things had apparently been moved. The complete report of this investigation of 2006/2007 is published in "The Hidden Chapter."

Keeping in mind that major alterations had been made at St. Michael's in 1846, the ongoing continuity of this religious site remains uncertain. Earlier G.P.R. by a different but reliable G.P.R. company, again commissioned by me, had demonstrated similar results in a nearby area of the churchyard. An example is the commemorative headstone I identified earlier in connection with Francis Bacon's burial. The commemorative stone is in place, but what 'remains' are below, is uncertain. My rationale has to consider whether decisions have been made officially from time to time, by those in a position to do so, to remove connections to a past no longer considered appropriate to the area. In other words, have Merovingian relics been returned to a base considered to be more appropriate? This could account for the 1850 visit to the area by Razé and the visit of King Charles I's wife, Queen Henrietta to Suzanna, Shakespeare's daughter. Both were French, with links to Tintern. Is it possible that precious items were transferred to St. Denis's in Paris? The Abbot Suger vase positioned strategically in the Tintern 'allegory' suggests that this might have happened. I have always thought that the vase was a major clue positioned as it is opposite the fish-like figure of Merovee. Speculation, maybe, but sometimes 'truth is stranger than fiction.'

Henry Somerset, 8th Duke of Beaufort had been born in Paris in 1824, the year of Razé's arrival in Canterbury. The Somerset family, together with the Archbishops of Canterbury would likely have been major influences on the structural changes in 1846 at St. Michael's Church. Traditions and family loyalties certainly affected the modernisation of roads, rail and river usage, which affected the landscape of Tintern in many ways. With the progress of time, those changes inevitably cause blurs in the memory as families move on in the developing world.

By 1878 The Nurtons was owned by the Reverend Clement Crutwell, but he did not become the resident incumbent of St. Michael's. Although some of his children were born at The Nurtons and the Reverend remained the owner for forty-one years, he obtained a 'living' at Frankby in Cheshire from 1889-1918. He spent his last years at Tintern, dying in 1929. During his absent years The Nurtons was leased to tenants.

It was from 1912-1918 that The Nurtons was the home of Oliver William Foster Lodge, the eldest son of Sir Oliver Lodge. Lodge junior, considered medically unfit to join the military forces in World War I, was a very talented architect, poet and artist. The Tintern allegory 'Stem of Jesse' discussed in chapter 12, I believe was created by him. Although admittedly I have not one hundred percent proof, I have been persuaded to that conclusion by a volume of one hundred and thirty poems he wrote and originally published in 1915 while he was resident at The Nurtons [6]. The volume is dedicated to his wife 'Wynlayne.' It becomes apparent from the structure of the work as a whole and the order of the poems, that the author was a very sensitive individual with high principles and very knowledgeable of history.

Some of the poems imply an in-depth knowledge and understanding of the historic associations of The Nurtons going back in time for centuries, with hints connecting it to the secret society 'Prieure de Sion,' said to be based in the vicinity of Rennes-le-Chateau, France. Poem '121' is a telling example. I include the title with the suggestive first six lines:

Cynara at Carcassonne.
Who sleeps under this dark stone?
I will tell you, such an one
As cold boys when she did pass
Pressed their foreheads to the glass,
And at night when in their bed
Unto her their prayers they said.

Carcassonne is but a few miles from Rennes-le-Chateau and is associated with legends arising in the area. Cynara is a thistle-like plant of the sunflower family, related to the artichoke, which was introduced into

Henry VIII's garden in 1530. King Henry considered himself to be a descendant of the Merovingian bloodline, a strong component of the mystery surrounding both places.

I include another poem from this collection because I think it reflects knowledge of a close-kept secret in its time, made evident in what I believe is his 'Stem of Jesse' painting. The number of the poem is '100,' centum, C, the Roman numeral associated with the commemorative headstone of Francis Bacon in St. Michael's graveyard. If Tidder Dee was Francis Bacon's son, christened Theodore who died on 9 April 1601 suddenly, could the poem be the poet's explanation of what might have happened? Lodge, the artist shows a youth at the top of the Pillar in the painting. Francis Bacon himself saw his own life as a deceit. I quote from Lodge:

> IRREVOCABLE
> She spake, he had to die. I took
> My friend into the little wood
> Where we had often played of old;
> We fought but I was armed, I strook
> My knife into his heart, the blood,
> Spattered my guilty arms. I left him cold.
> That was last year. This year she said,
> "Where is that pretty boy of yours
> Who used to sing me songs, and vow
> He'd love me while life lasted?" "Dead,"
> I told her. "Death's sweet cup that cures
> All hearts, hath his. We must not weep
> him now.

A sad poem with a mysterious certainty. The title 'Irrevocable' – 'unchangeable' says it all. That it is poem '100' is an echo of that certainty.

Oliver Lodge's time at The Nurtons suggests historic reflections, with a deep attachment. His son Oliver Raymond Wynlayne, whose mother died in childbirth in 1922, on visiting The Nurtons years later as a then-retired judge, said that of his parent's recollections of their married lives together "those at 'The Nurtons' were their happiest." In dedicating his collection of poems to Wynlane I will always be prepared to consider whether Lodge was recording his own comments in a literary way alongside Shakespeare's '154' Sonnet, published in 1609, a few months after John Dee's death in 1608.

The Sonnets were dedicated to 'Mr. W.H.,' in all probability William Herbert, 3rd Earl of Pembroke, patron of St. Michael's, who died after his celebratory 50th birthday dinner on 9 April 1630, at Fisher's Folly, London. Without further comment, I include:

Sonnet
153
Cupid laid by his brand, and fell asleep.
A maid of Dian's this advantage found,
And his love-kindling fire did quickly steep
In a cold valley-fountain of that ground;
Which borrow'd from his holy fire of Love
A dateless lively heat, still to endure,
And grew a seething bath, which yet men prove
Against strange maladies a sovereign cure.
But at my mistress' eye Love's brand new-fired,
The boy for trial needs would touch my breast;
I, sick withal, the help of bath desired,
And thither lied, a sad distemper'd guest,
But found no cure: the bath for my help lies
Where Cupid got new fire – my mistress' eyes.

The next Sonnet, 'no 154' is the final Sonnet in Shakespeare's collection, the first two lines of which are:

The little Love-god lying once asleep
Laid by his side his heart-inflaming brand,

Oliver Lodge moved to Painswick in 1918. The Reverend Clement Alfred William Crutwell retired from his Frankby living and moved back to The Nurtons, spending his last years there. He died on 31 December 1929. I had been monitoring the occupancy of The Nurtons for some time, so I was interested to learn who was to follow the Crutwells. Following almost immediately, a farmer from Narberth in North Wales purchased The Nurtons and moved to take up residence there with his family in 1930. His name was George Lewis James. The family remained there until 1959, almost thirty years.

By now I had become curious about anyone who was living at this address, so I looked into the background of the James family via the censuses. George Lewis James was fifty-one years old when he moved to The Nurtons. He had a twin-brother called Benjamin and was one of seven children. What became of interest to me was that he had an older brother called Evan Tudor James who lived with them, who died in 1954 and was buried at St. Michael's Church. I was bemused that his parents should have christened their eldest son with the name of Evan Tudor in 1877. The name 'Tudor' had already been a focus of my research for several years. This family had moved from North Wales.

The James family had been very well-established farmers in the Narberth area. Narberth Castle was built by Andrew Perrot in the thirteenth century.

Perrot's descendants held prominent Court positions through the centuries, particularly during Henry VIII's reign. Sir John Perrot served as Lord Deputy to Queen Elizabeth I during the Tudor conquest of Ireland. He died in custody in the Tower of London, having been convicted of high treason during that appointment. It was speculated by his contemporaries that he was the illegitimate son of Henry VIII. This was disputed, although he has been described as being of a similar appearance and temperament. Were the James family in some way associated with a legacy of Sir John Perrot 1528-1592 and Henry VIII? Hence the name Tudor as the middle name, with Evan as the first. Henry VIII had very close connections with The Nurtons and St. Michael's Church throughout his reign.

George Lewis James sold The Nurtons in 1959. His son Clifford John James had married Bethwyn Thomas, a nurse, who went to live in Australia. The family tradition is that she burnt many of the historic papers on-site at the The Nurtons, including the 'Visitors Book.' I wondered how many years that would have covered and whose signatures would have been there. This same George Lewis James, who bought The Nurtons in 1929, died on 7 October 1962 in a residential home in Worcester. The Probate of his Estate was granted on 28 May 1963.

In 1959 The Nurtons was unoccupied for a time, followed by short-term owners until 1963 when negotiations began for its sale again, under the auspices of the solicitor Francis & Co. of Chepstow. The next owner of the property was Harold Wood who took over the house and lands in 1964. The family remains the current owner.

The Nurtons, I believe, has been linked through the centuries with historic events of continuity and then change. It could be regarded as another coincidence, but I consider that it was a very controlled event when the property was eventually sold in 1963, the end of an era. In that year on 11 October, Jean Cocteau died. He had been ailing for some months and had been Grandmaster of the 'Prieure-de-Sion' for forty-five years by then. On the commemorative tableau behind his final resting place he had painted the image of Sir Percy Loraine in the likeness of Christ. Percy Loraine was the nephew of John Baldwin, friend of the 8th Duke of Beaufort. Baldwin, as already stated is buried in St. Michael's Churchyard. The chapter came to a close.

With the death of Jean Cocteau, the 'Prieure-de-Sion' it seems was changed as an organization. Irreplaceable documents in connection with local involvement in Tintern could have been destroyed at The Nurtons. Meanwhile, other major changes had taken place; the officially recorded Grandmaster was dead. I believe 1963 and 1964 may well have played a part in the pageantry of change for Tintern as well, for the role it played in the heritage of what had been known as the 'Prieure-de-Sion.'

L. L. Razé's poignant, single visit may well be a key to some of the critical responsibilities this beautiful area has kept close to its heart.

CHAPTER FIFTEEN

The Plan Preserved

O<small>N</small> 14 M<small>AY</small> 1992, the official launch of "The Byrom Collection" took place at 'Groucho's Club' in Soho, London. Arranged by Jonathan Cape and my literary agency, Andrew Mann, it was considered an appropriate setting for the book to begin its journey onto the world stage. The luncheon was attended by representatives of English Heritage, the publishing world, the media, journalists, as well as book reviewers and contributors to the book itself. So it was that Tom Maschler, the commissioning Editor at Cape, was rubbing shoulders with the current TV presenter Adrian Charles. Sir Miles Huntington-Whitely, Prime Minister Stanley Baldwin's grandson, could be seen talking about the book to a representative of the Trustees of the Byrom Collection on which the book is based. It was a glamorous affair. For me, the high spot was the speech made by my very experienced and respected Editor, Tony Colwell. I recorded a little of it in my daily diary; after all, I had considered him to be a hard taskmaster and I appreciated what he said:[1]

> Cape has never done a book like this before and I have never known an author so careful and meticulous in the preparation of the material without being oppressive ...

So be it. After the luncheon at Groucho's I completed a book signing at Hatchard's, a London book shop. For the next two days it was the usual round of BBC interviews, including an interview with John Dunn for the BBC Radio Programme 'Today.' I later came to understand that this interview was heard by Lord Michael Birkett.

Over the next weeks reviews of my book came out. It seemed to be taking its place as a positive piece of cultural research. The Reverend Neville Barker Cryer, having studied my research prior to publication, had made Masonic contributions where appropriate. He decided to write a review for Masonic research scholars. The Ars Quatour Coronatum no. 2076 is a Masonic Lodge in London dedicated to Masonic research. The Lodge meets at Freemasons' Hall, Great Queen Street. Some Masons dissatisfied with the way the history of Freemasonry had been reported in the past founded the Lodge in 1884. It began what is now called the 'Authentic School' of Masonic research.

In addition to quarterly meetings where papers are delivered and presenters questioned, the Lodge publishes yearly reviews titled 'Ars Quatuor Coronatorum' and maintains the Quatuor Coronati Correspondence Circle which allows participation from Masons all over the world. The Reverend Mr. Cryer as Secretary wrote a review which was published in the October edition of the review later that year. I think that his Mason-inspired comments are consistent with our understanding and I include them here as published in the article:[2]

For long enough the cry has gone up in masonic study circles, 'Oh, if only there were some new material to work on'. In the appearance of this book, as well as in the biography of John Byrom that is shortly to follow it, I believe that we have something as an answer to that cry. Brought by the strange and unexpected history of those associated with her own home in Manchester to a study of Byrom, Joy Hancox was launched on a detective trail that brought her into contact with areas of knowledge, including Freemasonry, with which she was completely unacquainted. She herself is still amazed at where she has been led by the evidence of her researches. The surprise and the excitement come over in the narrative. Though the full facts must await the appearance of Byrom's biography, the reader of this book can gather that much of that man's diary printed in the Chetham Record Series was omitted because it was recorded in a shorthand that he devised. Indeed, it was Byrom's shorthand method which was later purchased by a certain Isaac Pitman and turned into the method now so familiar. By learning the Byrom technique Joy Hancox was able to unravel events in the life of this author of 'Christians awake, salute the happy morn' which had hitherto remained hidden and private. One of them was that Byrom was a freemason in the earliest days of the Grand Lodge and had links with many previously known to us there and in the Royal Society. His own 'Cabala Club' was held on the lower floor of the meeting place of the Lodge of Antiquity and frequently at the same moment. Hans Sloane once invited Byrom to come up and join them.

It is therefore in the same setting as that which surrounds the emergence of 'Grand Lodge' Masonry that Joy Hancox found herself having to unravel a 'find' that was in the possession of Byrom's only remaining descendants. This was a 'Collection' of 516 drawings which by their texture, watermarks and occasional notations have proved to be a gathered selection of architectural, nautical, symbolic and even cabalistic representations from the late sixteenth to the early eighteenth century. How they came to be in Byrom's possession is part of the story here revealed but even more exciting is the manner in which, by determined detective work, Joy Hancox has unravelled the 'secrets' which these drawings retained.

Using her training as a professional musician and teacher she has recognised the 'proportions' and 'harmonies' that these very detailed drawings portrayed, and when to this discernment is added the knowledge which she acquired of the earlier devisers of these pieces, then the Collection can be seen to have more than usual importance in several fields of study. Their implications for the construction of the South Bank theatres such as the Globe and the Fortune, and others, are fully explained whilst the link of other drawings with the Temple Church and other Templar buildings, or the careful and meaningful construction of King's College Chapel, suggest that here is some explanation for that recurrent masonic belief that this Craft did have 'secrets and mysteries' of real value. The assertion of Bro. Haffner that 'architecture' was a subject of real importance in early lodges may not have been as wide of the mark as many of us seemed to think. What now seems clear is that there was a tradition of 'design knowledge' that persisted *after* the Reformation and was both preserved and handed down to groups of gentlemen students.

This is a book that has been most carefully prepared and beautifully produced, as its price reflects. Here is no sensationalism or shallow presupposition. Evidence is produced for each step taken and the conclusions are only those related to the facts available. There is here ample ground for much further examination by masonic students and the following biography will provide further material for extending our 'search'. Joy Hancox does not attempt to make any suggestions about the implications for Freemasonry as that is, she readily admits, not her field nor her principal interest. The book has enough fascination for the reader in the main revelations that it proffers, but its pointers towards Freemasonry are also not hidden. Here at least is a source of new thinking and evidence that we shall ignore to our loss. To put it at its lowest it is a very good read.

* * *

To my utter surprise, within a few months of the publication of volume 104 of 'The Premier Lodge of Masonic Research – no. 2076,' I received a letter requesting a meeting from 'Bro. The Revd. William Stemper.' In keeping with Masonic tradition, an article of his had appeared in the same Volume (104). He was an American who's office he gave as The Rockefeller Center, New York, New York. The Revd. Stemper was requesting a meeting to discuss 'Byrom.' The title of his own published article which prompted him to write to me was 'Conflicts and Development in Eighteenth-century Freemasonry: the American Context.'

The introductory lines of this article will provide, in principle, an understanding of the research which was being carried on by the Reverend at the time:[3]

The Development of Freemasonry in the eighteenth century is a complex and tangled story. It has not as yet been unravelled in terms of the origin, development, and theological significance of its symbolism synthesis, and subsequent development within English Freemasonry, which generated a unique masonic tradition in America.

The Revd. Stemper was the Bishop's Vicar for Corporate Affairs of the Episcopal Diocese of New York, U.S.A.

At the time of his contacting me he was completing his Doctor of Philosophy degree at Oxford University. Already a fellow of the Royal Society of Arts, London, he suggested that we should have our meeting at their headquarters, at the same time enjoying tea in their club rooms. Unfortunately, it was not possible for me to make the journey in the suggested time frame available, so I wrote to Revd. Stemper, inviting him to Manchester to study Byrom's Research papers. I include his response:[4]

Many thanks for your gracious note, re. John Byrom, received yesterday (8 July). It was a privilege to hear from you. My research and writing duties make it difficult to leave Oxford in the short term, but I would very much like to make a visit, as it becomes more possible, to Manchester where I have never been. I look forward to studying your excellent book(s) in more detail, and to a future of a pleasurable meeting, along mutual interests.
Yours faithfully,
Bill Stemper.
9 July 1993.

I include Revd. Stemper's letter, along with the review of "The Byrom Collection" by Neville Barker Cryer, as a reflection of the interest expressed by scholarly Freemasons of international standing. Unfortunately, Revd. Stemper had to return to America in the August. We were unable to meet as we had hoped. His interest continued. He died in 2009. So it goes on.

During the occasion of the book launch at Groucho's, in acknowledging its reception, I said to those present that the book would now be making its own journey. As long as there was one copy left on a book shop shelf, the knowledge was available to be shared. At the front of the book, I had included a quote from Francis Bacon himself:

Whatever deserves to exist deserves also to be known, for knowledge is the image of existence...

This philosophical statement was included by Bacon in one of his own publications of the early seventeenth century. I had included the quote in my book without realising at the time how close he would be to the story of my own research some four hundred years later. His contribution to the world of science and subsequent developing technology could be such that the ideology of his statement would become outmoded and not necessarily true.

Having included the Rev. Mr. Cryer's review in full and a quote from the Rev. Stemper's article, both from the annual review of the Research Lodge of Freemasonry, I came to realise over the years that some knowledge was not freely available for all to access. My years of concentrated study and the search for information which would account for some of the conclusions I'd arrived at, persuaded me to include in my previous book the results of an avenue of research in connection with The Nurtons. It had to do with a diary entry under the heading 'Noah's Ark,' which when investigated, I learned was associated with Freemasonry. I quote from "The Hidden Chapter":[5]

> When the foundation stone of the new Silurian Masonic Lodge was laid in Newport on 29th August 1855, the Lodge members proceeded ceremonially through the streets to St. Paul's church and afterwards to the Town Hall. The proceedings began in the open air before a large crowd with these words: "Men, women and children here assembled today to behold this ceremony, Know all of you that we be lawful Masons, true to the laws of our country and established of old with peace and honour in most countries, to do good to our Brethren, to build great Buildings and to fear God Who is the great Architect of all things. We have among us, concealed from the eyes of all men, secrets which may not be revealed, and which no man has discovered, but these secrets are lawful and honourable to know by Masons who only have the keeping of them to the end of time."

Many of those listening on that occasion would not have been Freemasons and one wonders whether some might have felt intimidated by that statement. Freemasonry claims that it is not a secret society. But clearly at that time, it was a society with secrets and those secrets were solely for the use of those within the brotherhood. I, therefore, would not have been eligible to know what those secrets could have been as declared on that day in August 1855. The occasion was just a few weeks after the entry concerning 'Noah's Ark' in Jenny Audland's diary for 12th March 1855. Noah, according to Masonic legend, discovered 'the stone of foundation' and placed it in the Ark as an altar. My interpretation of Jenny's enigmatic brevity is that she was recording a meeting at The Nurtons on that date for members of this exclusive organisation. (Jenny was the wife of John Audland, the general practitioner in Tintern). By 1855 plans were already under way to build a new Masonic Hall in Newport, Wales:

> The plans (for the new lodge) were submitted in February, £870 had been collected by March, and by June the P.G.M. (Provincial Grand Master,

Charles John Kemys-Tynte) had agreed to attend the Dedication ceremony of the stone laying.

John Byrom was a Freemason of course, but he never mentioned his membership in a Lodge in his diaries or published poems. The fact that he owned this collection of geometric drawings demonstrates his ability to maintain confidentiality where, when and as required. The geometric drawings had slipped through any identifiable net from his death in 1763 until they were sent to me at my request in 1984, in two brown paper bags.

In all Masonic Lodges, geometry as a discipline was and I assume still is paramount within the organisation. The letter 'G' is to be found in the centre of the ceiling of most Lodges. In American Lodges it is to be found on or near to the Master of the Lodge's chair. It is also acknowledged that 'G' as for God stands for the Great Architect of the Universe. In some foreign Lodges, there has been a misunderstanding of the symbol. Sometimes a triangle is used for containing the Hebrew name for God.

For Freemasons, geometry is regarded as the root and foundation of all arts and sciences. The rule of measure, together with the associated lines and angles, dictates how geometry is used. How various civilisations throughout the centuries have applied it differs within cultures. The variables between and amongst the differing disciplines can cause confusion. Yet, the beauty of number is a certainty upon which clarity of purpose creates confidence. Number can be applied to those disciplines using multiple systems and symbolism.

Having recognised an alternative measure within the collection of pieces of paper and card in the Byrom Collection, I was able to use numbers in different formats for different disciplines, for example the architecture of buildings and even for philosophy. Over the years there were difficulties I encountered where specialists with differing expertise interpreted number symbolism in different ways. Sometimes it is a matter of opinion. As I discovered, the term 'sacred geometry' carries with it its own agenda.

The Rev. Mr. Cryer and I had meetings for over twenty-three years. They were always purposeful with an immediate agenda, which itself had a specific aim – each of us was to learn from the other. Where my pursuits for new information had taken me and with whatever results, interpreting and gathering the facts associated with the drawings was always the prime factor. The Reverend had made it clear to me right from the start that he had no time to waste and did not spend time at meetings where there was not likely to be of some benefit and usefulness for him; I was in like fashion. I learned a lot from him over the years. Even so, I always felt that the vows taken within the Freemasonry organisation inhibited a completely open dialogue between us. I accepted that.

There was much I learned over the years that I would not have become aware of if I had not had the long-term acquaintance with the Rev. Mr. Cryer. Membership of the social aristocracy in the Lodges throughout the centuries guaranteed that a measure of control of much in the way of culture and developing innovation remained within their ranks. Exclusivity was their 'by-word,' I was always mindful of the statement made outside the Silurian Masonic lodge on 29 August 1855. At the same time, the Reverend's visits over the years suggested that any contribution I made during the meetings was valuable to the organisation, if not unique. There was a mutual respect.

Separately, I came to realise that when making enquiries through public bodies or individuals with no masonic connection, responses were often innocently naïve or non-committal. That, however, was not the case with students of sacred geometry. Usually very scholarly and dedicated individuals, they had to come to terms with the fact that some of the drawings within the Collection were so personalised to the aims of selected members of the groups Byrom associated with that their application to the principles of sacred geometry could not be interpreted accurately unless one had access to the original measure used in their creation.

The 'rationale' or basis for that measure was different to those that were by then accepted principles, as practiced by many. To be accurate, it can best be described as 'similar' but 'not the same.' Sometimes this led to a degree of bewilderment with certain authorities wondering whether I had the ability to deal with such high-minded information. Time and chance helped to resolve some of these issues, but clearly not all.

I recall an instance in May 2009 when a correspondent drew my attention to a book entitled "Dark Mission: The Secret history of NASA" by Richard C. Hoagland and Mike Bara. It had been published in 2007. The book has aroused a great deal of controversy from claims made by the authors regarding secret 'brotherhoods which quietly dominate NASA,' with astronauts being involved, such as the very senior Freemasons who allegedly took a Masonic flag to the moon in 1969 during the Apollo 11 mission.

I have no intention of discussing the rights or wrongs of that publication, except that the reason I was told of it was that there was a geometric figure represented in it which replicated, in principle, John Dee's 'Monad,' of which there are also representative copies in the geometric collection of John Byrom. I had reproduced them already in my books. Was there a relevance? The Apollo moon visit had taken place years before I had commenced any work on the collection. The NASA book also makes a reference to John Dee and Edward Kelly 'communicating with angels' using the 'Enochian Code language.' This reference was used to imply that NASA astronauts, engineers, program directors and others closely connected to the program were familiar with John Dee's records in their considerations of astronomy and mathematics (and even astrology!) and that there is a secret cabal dominating the agenda of NASA. Perhaps. Who knows?

Because of the connection with Freemasonry in that book, I asked the Rev. Mr. Cryer on his next visit whether he was aware of it. He said he was not but ordered a copy of the book from a local book shop. Reverend Mr. Cryer knew that I was anxious to make accurate contributions through my research. So quite independently in September 2008 he said he would write to a Masonic colleague in America with an enquiry on my behalf regarding the 'Apollo 11' Masonic Lodge flag and the John Dee reference as quoted in "Dark Mission." In the meantime, he read the book himself.

The Masonic colleague he chose to write to was Bro. S. Brent Morris and eventually he received a reply dated 10 June 2009. We had no meeting arranged around that time so the Reverend Mr. Cryer forwarded the complete seven pages of the correspondence sent to him. Brother Brent Morris is a Master Mason, a 33 degree Scottish Rite Mason and editor of 'The Scottish Rite Journal,' a publication of the Supreme Council of the Scottish Rite Southern Jurisdiction. The letter was written on Masonic-headed note paper. [6] In his reply Brother Brent Morris said that astronauts are allowed to take a limited amount of personal material with them in their flights and three astronauts took Masonic flags. He gave the names as:

1965   Leroy Gordon Cooper – Gemini V
1968   Walter M. Schirra – Apollo 7
1969   Buzz Aldrin – Apollo 11

The flags were brought back and are in the Museum of the Supreme Council (U.S.A.). The flag shown on the cover of the book 'Dark Mission' was taken to the moon but the illustration on the cover had replaced the USA flag or as they say, been 'photo-shopped.' On 9 July, I replied to Rev. Cryer and include my response to Brother Morris's explanation:

Dear Neville,
Thank you again for forwarding the summer edition of 'Freemasonry Today'. As you will imagine I value having the copy very much. I also appreciate having a copy of S.B. Morris's letter. My interests and concerns regarding the Hoagland book have been validated I feel, in that several of those awesome astronauts are indeed members of Sc.Ri. and I am glad to know that for various reasons. That Hoagland's (and Bara's) illustrator 'doodled' with the flag photographs was in my opinion very misguided,

giving the opportunity for other questions to be raised about validity, where perhaps their researches are intelligent and interesting in some respects...

Three astronauts, all 33 degree Masons, had each taken their Masonic flag to the moon on separate missions. What does it mean to be a 33 degree Mason, one might ask? The general definition is as follows:

> In the United States, members of the Scottish Rite can be elected to receive the 33 degree by the Supreme Council. It is conferred on members who have made major contributions to society or to Masonry in general.

Their achievements were awesome. By coincidence Buzz Aldrin, the Apollo 11 'moon-walker' astronaut published his autobiography "Magnificent Desolation" in 2009, the same year the response from Brother Brent Morris was received. Full of admiration, I bought a copy. Would there be any mention of his membership of the 'Brotherhood' or the flag episode? I have no hesitation quoting his own words describing the occasion, after all he was one of the first human beings to have walked on the moon, along with Neil Armstrong:[7]

> Just a short distance away from the LM, we found a spot to put up the American flag. But getting the flagpole to stand in the lunar surface was more difficult than we anticipated. The pole itself was hollow, and we were trying to push it down into the lunar 'soil' that was made of millions of years-worth of asteroid impacts, all densely packed down into a hard surface.

Buzz Aldrin's personal note is vivid. There is no mention of the Masonic Lodge flag and nowhere in the book does he say that he had been elected by the Supreme Council to be a 33 degree Mason.

So, what was I to make of the reference in "Dark Mission" to John Dee and Edward Kelly? As I said, it was the reference to the two men "communicating with angels in the 1500's using the Enochian Calls." One of the aims was to create a 'moon child' who would open an 'inter-dimensional doorway' to outer space. Although I have had to investigate aspects of what could be described as the 'arcane world' in my research, it had always been with the purpose of making enquiry and learning. At the same time, it must be said that the Collection covers many disciplines of higher learning. It is not limited to what we now consider 'appropriate.' At the time of my acquiring "Dark Mission," I knew nothing of those 'occult' practices, as more recently explored by John Whiteside 'Jack' Parsons in America in the 1940s. It was the diagram of the symbol for 'Mercury' reproduced in the book, which I placed alongside Dee's 'Monad,' which had aroused my curiosity.

As described earlier, Edward Kelly and John Dee formed their agreed pact (together with their wives), having had a consultation with the Enochian Angel Madimi. As a result of this agreement, Dee's wife Jane gave birth to Theodorus Trabonianus (Tidder Dee) on 28 February 1588. I continue to hold the opinion that he was Edward Kelly's son (aka Francis Bacon's son). By then I had studied the Fenton Edition of John Dee's diary very carefully.

Since becoming the licensee of the Collection, I gradually developed reservations as to whether a facsimile of the complete collection should be made available during the time it was under my contract and control. That had been an early ambition of mine but I learned that some of the uses of the mathematics inherent in certain drawings possessed negative forms of energy. I recognised that fact as a responsibility in making any decision about availability. My concerns regarding the Hoagland and Bara, and Dee and Kelly quote took on a different level of meaning. I had already deduced that several of the drawings as represented had something to do with 'propulsion,' call it a hunch. John (Jack) Whiteside Parsons's legacy to Science concerns 'rocket propulsion.' Besides his more esoteric pursuits, he was also an engineer who worked on early rocket engines and designed and built rockets.

A colleague of Parsons in his 'occult' workings was the sinister Aleister Crowley, who had it seems a connection with the owner of a large estate in Monmouthshire in Wales, that of Evan Morgan, 2nd Viscount Tredegar of Tredegar Park. The Viscount claimed Crowley as a 'friend and admirer.' The 13th Duke of Bedford, also with connections to the Tintern story, described the Tredegar family as "the oddest family I have ever met."

Evan Morgan succeeded to the title of Viscount in 1934 on the death of his father. Morgan was deeply involved with the occult and occult practices. Aleister Crowley described Morgan as an 'Adept of Adepts.' The 2nd Viscount Tredegar, Evan Morgan, could claim another distinction. Near to his estate, Tredegar Park, during his confinement, was Rudolph Hess, Adolph Hitler's deputy Fuhrer before and during the war. Hess was a contemporary of Evan Morgan and Hess himself had deep family ties to the area. Hess also, it may be noted, had an interest in the occult, an interest he shared with Adolf Hitler. Hess is said to have declared that he had conceived of his ill-fated 'peace mission' to Britain in a dream, which Winston Churchill described as an act of "lunatic benevolence." The point to be made is that it was under Hitler's regime that the first 'rockets' were built. The technology was very advanced for the time.

Having crashed his plane in Scotland on 10 May 1941, Hess spent some months in confinement in Aldershot before being moved to Maincliff Court, near Abergavenny in Wales, on 26 June 1942. He remained there until his trial in Nuremburg in 1946. Maincliff Court was an Asylum hospital known as Maincliff Court Military Hospital and POW Reception Centre.

The 2ⁿᵈ Viscount Tredegar's family home for over five hundred years, Tredegar House, was situated just half-hour's drive away, near Newport. Rudolph Hess's father, (Johan) Fritz Hess, who had lived in Britain for many years, had married a local girl from near Newport, Elizabeth Mackie, on 28ᵗʰ October 1890. They were married at the church in the parish of St. John, Clerkenwell, London. Her death certificate stated that she died aged thirty-five years of a kidney disease during her first pregnancy (shortly after the marriage) on 13ᵗʰ June 1891. In 1890 she had met 'Carl' Fritz Hess who worked in the Bishop of Gloucester's Palace as a steward and/ or Butler, a position requiring confidence and trust. On their marriage certificate he is described as a 'Butler.' Calling himself Carl Hess, he had previously worked in London for some years as a mercantile clerk.

After the sad loss of his wife Hess returned to Hamburg, Germany. He remarried and went to Alexandria in Egypt, having inherited a family business there. Rudolph was born in Alexandria on 26 April 1894. Since Rudolph Hess lived at the Maincliff Hospital for three years during and after WWII, which was situated in the area where his father had close connections, it seems possible that son Rudolph, who shared interests and his own connections to the area might have met Evan Morgan during his stay under confinement.

The lifestyle of the 2ⁿᵈ Viscount Tredegar, Evan, has been described as extravagant to such an extent that the family fortunes were adversely affected. After five hundred years, in 1951, the Morgan estate was sold. Evan Morgan had no children of his own. Rudolph Hess, at his trial in Nuremburg in 1946 was committed to Spandau Jail near Berlin. He remained there for the rest of his life, dying in 1987 at the age of ninety-three, by suicide, hanging himself with an electric cord in his cell.

* * *

Over three hundred years earlier, Francis Godwin, the Bishop of Llandaff, Cardiff, not far from the scene at Tredegar, had written a book called "The Man in the Moone." Described in popular culture as the first science-fiction of its kind, it describes how a Spaniard, Domingo Gonsales, is ferried to the moon by a team of Swans meeting the 'Lunars' resident there, a gentle non-violent race. Not the same colour as humans, they were blue/green. Godwin is said to have borrowed the concept of trained birds from Francis Bacon's 'Sylva Sylvarum' (Natural History.) Bacon's book was published in 1627 (one year after his supposed death). Godwin's book was published in 1638. Bishop Godwin himself died in 1633.

As the Bishop of Llandaff, Godwin and his family lived at 'Moyne's Court,' Mathern, near Chepstow, not far from Tintern, from 1601 to 1617. That Bishop Godwin knew Francis Bacon and John Dee is more than likely. As the resident Bishop of Mathern Church we find him writing to his friend William Camden on 27 May 1608. The letter included the following passage:[8]

> This name of Theodoricus putteth me in mind of Theodoricus rex and martyr, that lieth here in our church in Mathern, and gave unto the place the name Merthyr Tewdric...His tomb partly ruined. I have repaired, and added a memorial or epitaph, the copy wherof I send you enclosed.

William Camden was the tutor of Arthur Dee, John Dee's eldest son. John Dee himself died a few months later, on 26 February 1609. Tewdric had connections with Tintern and St. Michael's Church, already well documented. It is enough to note that he was revered as a descendent of the Merovingian dynasty.

Bishop Godwin's fertile imagination in his 'storyline' of "The Man in the Moone" became more curious to me when I remembered that he was a contemporary of another Bishop, William Overton, also a friend of John Dee. The Bishop died a few weeks before Godwin's letter to Camden on '9 April' 1609. As the Bishop of Litchfield, I believe he received and buried the remains of Theodorus (Tidder) Dee, who had also died on 9 April in Manchester, in 1601/2, according to which calendar is used.

Another professional member of the church, the Reverend Roger Williams, also a contemporary of Francis Bacon and referred to in the story of the Tintern allegory 'Stem of Jesse', later emigrated to America with his wife and was closely associated with establishing the Baptist Church in Rhode Island. The Archbishop of Canterbury, John Whitgift, confidante of Queen Elizabeth, taught Francis Bacon classics in the 1570's and was with Queen Elizabeth when she died in 1603. Whitgift crowned James I and received eight Bishops at the Hampton Court Conference in 1607, before dying himself later the same year.

Drawing these strands together, it was the 'Estate Owners' who from generation to generation inherited not only land and fortunes but 'family' knowledge. With time some of those connections and 'hidden knowledge' may have become exclusive and unwittingly secretive. As society became more technical and cosmopolitan, harnessing the facts of this scattered historical data demands a scrutiny of an unbiased kind. Not easy.

By the time I had considered the material drawn to my attention by the "Dark Mission" book, which I have recounted here, I was feeling it was time to pause. The miscellaneous strands of information about these individuals that I had gathered independently did not sit comfortably with me. In addition, I would be reminded from time to time of the unusual expertise that is required to ferret out the knowledge of this special sacred geometry and the language of number symbolism as used in different disciplines. As my research for this book progressed, the accumulation of data connected with the Rev. Neville Barker Cryer became a resource unto itself. Most of the materials used during our meetings was returned to me after his death in 2013.

In addition, some books from his own special collection and related investigations became part of that.

The gradual unfolding of the Tintern story in my own investigations was still not complete. I had not been able to identify what exactly had prompted the link with Nuremberg and the wireworks. The repetition of place names and people and their unusual interests came to seem less and less like a coincidence and more and more like some sort of 'divine' plan. My story began with the sixteenth century, upon my discovery of the hermetic and sometimes seeming occult arcane knowledge which had been recorded then. The latest link in the nearby area of Rudolph Hess's confinement was a troubling detail. How is that related?

The Byrom geometric drawings connected with 'propulsion,' as I saw them, became for a time a preoccupation. The quote in "Dark Mission" about John Dee and 'Edward' forced me to return to the topic. While studying the expertise needed for the Apollo missions, I encouraged myself to study the experts involved with master-minding space travel. It was inevitable that I would need to look at the achievements of Werner von Braun, the German-born American aerospace engineer and space architect. He was the leading figure in the development of rocket technology in Nazi Germany and a pioneer of rocket and space technology in the United States. Von Braun arrived in America in June 1945, his transfer and that of his specialist team of scientists having been approved by the U.S. Secretary of State. Von Braun joined the staff of NASA as a director in 1960 and remained in that position until 1970, retiring in 1972. He headed the Apollo Application Programme and died of pancreatic cancer in 1997 aged sixty-five.

The Apollo programme director Sam Phillips has been quoted as saying that the United States would not have reached the moon without the help of Von Braun. It was while studying the work of Von Braun on propulsion that I came across a diagram of the schematic 'design' for the original V2 rocket. Von Braun's interest in rockets had been originally for use in space travel, although rockets were launched towards England in 1944. By then Hitler knew that the war was lost. They did damage and killed people.

As I studied the principles of the schematic design and its layout, I was reminded of a similar layout in a geometric drawing from the Byrom

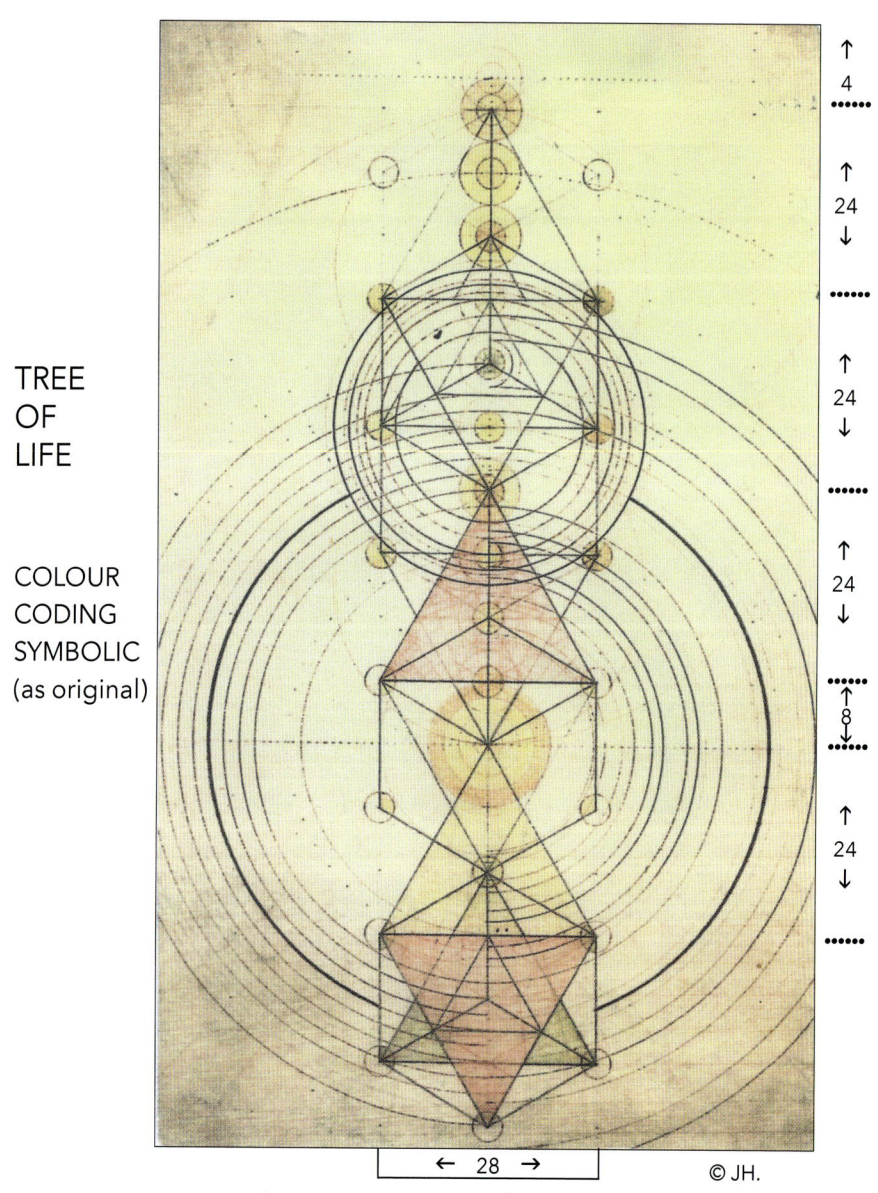

TREE
OF
LIFE

COLOUR
CODING
SYMBOLIC
(as original)

↑
4
••••••

↑
24
↓

••••••

↑
24
↓

••••••

↑
24
↓

••••••
↑
8
↓
••••••

↑
24
↓

••••••

← 28 →

© JH.

The Tree of Life 'The Byrom Collection'

SCHEMATIC – V2

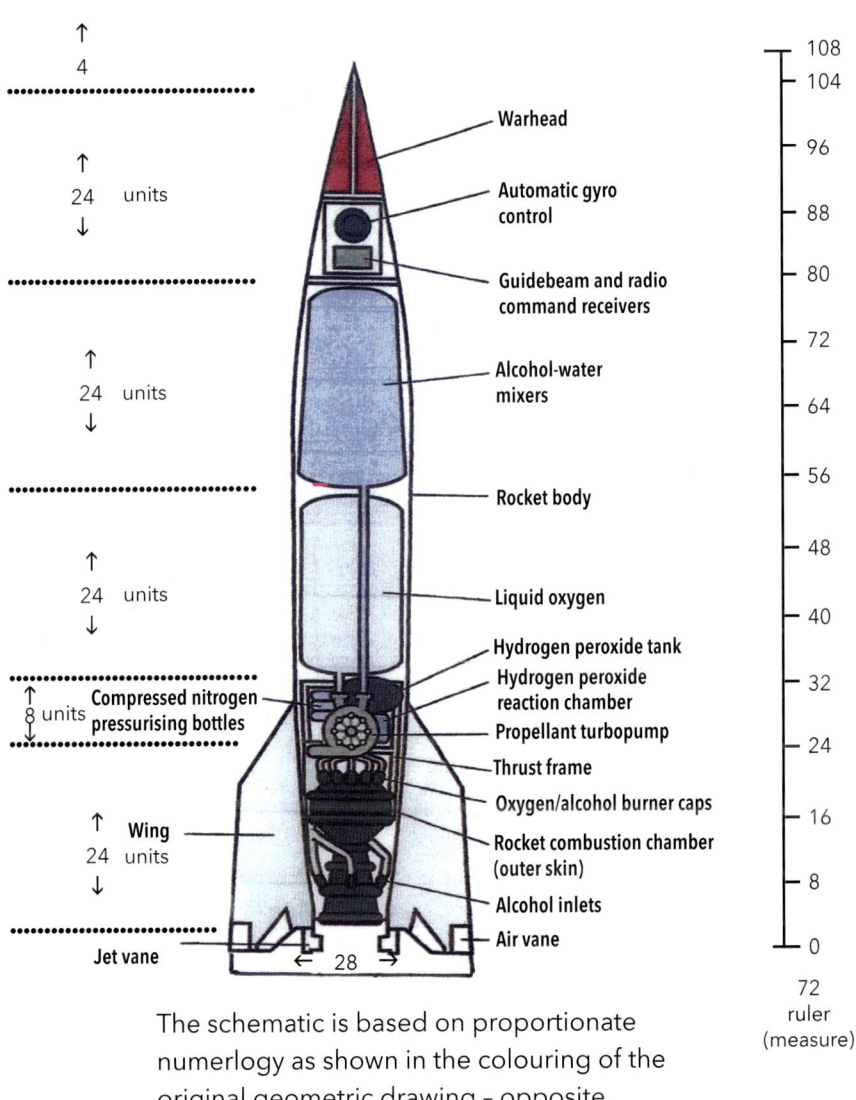

The schematic is based on proportionate numerlogy as shown in the colouring of the original geometric drawing - opposite.

© JH.

Schematic V2

SCHEMATIC - V2

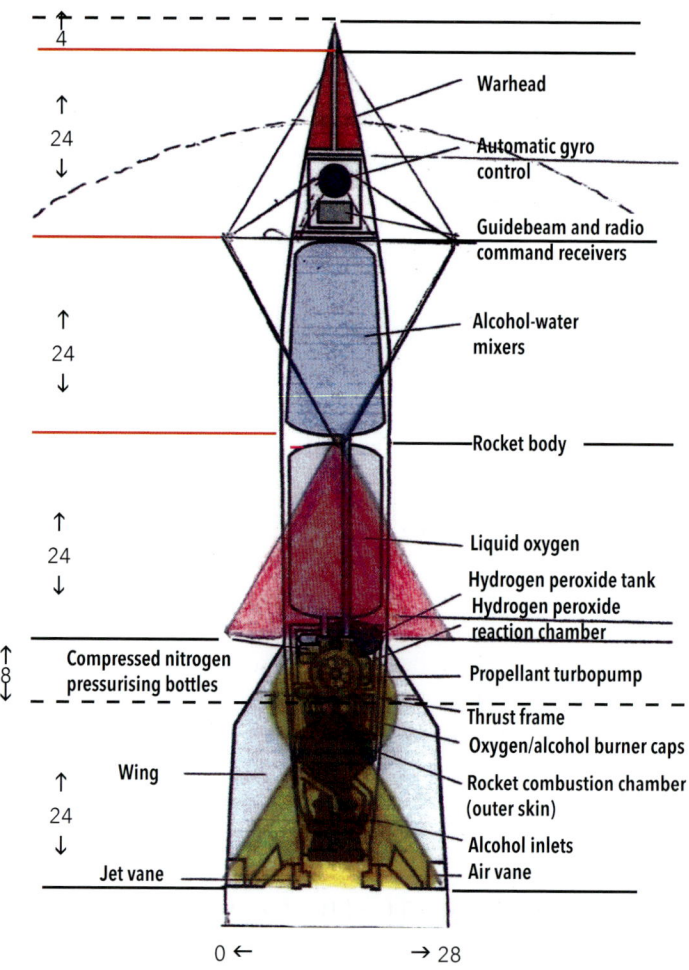

4

24

24

24

8

24

Warhead

Automatic gyro control

Guidebeam and radio command receivers

Alcohol-water mixers

Rocket body

Liquid oxygen

Hydrogen peroxide tank
Hydrogen peroxide reaction chamber

Compressed nitrogen pressurising bottles

Propellant turbopump

Thrust frame
Oxygen/alcohol burner caps

Wing

Rocket combustion chamber (outer skin)

Alcohol inlets

Jet vane

Air vane

0 ←   → 28

Compatible – the two together.
Colour coding features demonstrate how
numeracy and proportion seemed to have been
part of the decision making.

© JH.

Superimposed V2 and tree of life

234

collection. Searching through the catalogue I found it. To my amazement it was a drawing on the reverse side of another one that had appeared in my book "The Byrom Collection," published in 1992. In a practical sense the two drawings on that paper, front and back, were compatible and could be worked together. To my astonishment, using my '72' ruler, it became clear that Von Braun's 'schematic' for the V2 rocket could be matched measure for measure with the geometric drawing used in my book – catalogue no. 178 in the collection.

I set out a developing programme, with the premise that the Byrom drawing was one exclusively designed for propulsion. In fact, it replicated Von Braun's design schematic for the V2 exactly. Placing one on top of the other as their respective sizes were enlarged or reduced proved it. I was as startled by that match as I was when I was able to demonstrate how the Byrom drawings could be identified with the brass plates in the Science Museum using my '72' ruler. The matching geometric drawing from the Byrom collection is the one based on the complex structure of 'The Tree of Life,' a design linking the laws of nature with a philosophical 'ladder' moving upwards towards perfection, with the aim of guiding an individual's journey through his/her life on earth.

I knew where Byrom's collection had been housed since his death in 1763. The use of the '72' measure had seemed to be rare at the time. The fact that von Braun's schematic for the V2 could be matched in any way seemed uncanny to me and beyond my understanding. Where had such information been stored away? The only explanation I can offer at this stage is that John Dee's 'Monas Hieroglyphica' (Monad) had been dedicated to the Holy Roman Emperor Maximillion I and John Dee and Edward were familiar figures to his successor, his son Rudolph II and his court staff in the 1580's in Prague. Could the legacies of such knowledge have been deposited there for reference?

Knowing little of Von Braun's personal life I was struck by a statement he made in his own book "The Third Book of Words to Live By" published in 1962 by Simon & Schuster and I have decided to reproduce it here as a quote in full:

Today, more than ever before, our survival – yours and mine and our children's – depends on our adherence to ethical principles. Ethics alone will decide whether atomic energy will be an earthly blessing or the source of mankind's utter destruction. Where does the desire for ethical come from? What makes us want to be ethical? I believe there are two forces which move us. One is belief in a Last Judgement, when every one of us has to account for what we did with God's great gift of life on earth. The other is belief in an immortal soul, a soul which will cherish the award or suffer the penalty decreed in a Final Judgement. Belief in God and in immortality thus gives us the moral strength and the ethical guidance we need for virtually every action in our daily lives. In our modern world many people seem to feel that science has somehow made such 'religious ideas' untimely or old-fashioned. But I think science has a real surprise for the sceptics. Science, for instance, tells us that nothing in nature, not even the tiniest particle, can disappear without a trace. Think about that for a moment. Once you do, your thoughts about life will never be the same. Science has found that nothing can disappear without a trace. Nature does not know extinction. All it knows is transformation! Now if God applies this fundamental principle to the most minute and insignificant parts of His universe, doesn't it make sense to assume that He applies it also to the masterpiece of His creation – the human soul? I think it does. And everything science has taught me – and continues to teach me – strengthens my belief in the continuity of our spiritual existence after death. Nothing disappears without a trace.

Given the stature and life's experience of Werner von Braun, I choose not to comment on his own published statement. Readers will reflect on it in their own way. But I will return to the continuing disquiet surrounding Rudolph Hess's three-year confinement at 'Maincliff Court' near Newport and his father's earlier connections with the area. The newly learned data suggested by all these connections, proven visually by the geometric illustrations and the schematic of the V2 rocket, was of great significance.

It was time for discretion, pending further investigation. I did not have to wait long. In conversation with two of my trusted consultants, my attention was drawn to the content of a scientific thesis titled "Ether Technology" by Rho Sigma and published by him in 1977. The Preface begins with a quote by Sir Francis Bacon from his thesis "Novum Orgarnum" 1620. I think it appropriate to include the quotation below: [9]

> Those who have handled Sciences have been either men of experiment or men of dogma. The men of experiment are like the ant; they collect and use; the reasoners resemble spiders, who make cobwebs of their own substance. But the bee takes a middle course; it gathers its material from the flowers of the garden and of the field but transforms and digests it by a power of its own.

The study of 'Ether Technology' involves a principle in which astronauts are somehow followed into space and "tested as they changed due to high 'G' stress and other influences they were subjected to." It was all to do with 'Eloptic Energy' and 'light rays.' Scientific studies into radio-activity and related topics have been studied since the beginning of the 1900's but it was a paragraph on page 70 of this publication that gripped my attention: [10]

The incredible importance of research on this particular subject (ether radiation) had been recognised by Rudolph Hess, deputy of Adolph Hitler, who privately financed Dr. Joseph Wuest. This fact was mentioned to this writer by Dr. Wuest personally just a few years ago. Only World War II put an end to his research.

The writer here was 'Rho Sigma,' a pseudonym he used for himself, 'Rolf Schaffianke.' Described as a physicist and rocket scientist, he was the youngest member of the team who worked with Werner von Braun and moved with him to America in 1945. Engaged under contract to NASA as a consultant to the propulsion laboratories in Huntsville, Alabama, he died in 1994 aged seventy-three.

Rolf Schaffianke had worked closely alongside von Braun for years and chose to quote Francis Bacon at the beginning of his thesis 'Ether Technology.' That he was also aware of Rudolph Hess's patronage of Dr. Wuest and the nature of that support convinces me that Hess's confinement in 'Maincliff Court' was not simply an accident. It was the anonymous origin of the brass plates in the Science Museum and catalogued as the 'Boyle Collection' that had originally persuaded me to look into the history of the wireworks of Tintern in the 1990s.

As the years went by the mystique surrounding the area of Tintern grew while the breath-taking beauty verbalised by Wordsworth and others guaranteed a regular stream of sightseers to Tintern Abbey. Deep-seated questions surrounding that area through the centuries have been studied, and while the commitment of families associated with the legacy of the legendary William Shakespeare have been revealed in part, the hermetic and masonic links have been also been highlighted. This new link now provided by the matching of von Braun's 'schematic' of the rocket to the Byrom complex 'Tree of Life' took my findings to another level. I believe this 'Tree of Life' drawing to be the ultimate illustration of John Dee's beliefs regarding his 'Monad' philosophical teachings. William Herbert of St. Julian's and The Nurtons worked with John Dee on that treatise as well.

Francis Godwin, the author of 'The Man in the Moone' published in 1638 as Bishop of Llandaff, together with records surrounding The Nurtons and St. Michael's Church in Tintern, have brought the history of this area in Monmouthshire into a different realm of focus and intrigue. These were men of influence in their own day and their legacies will live on for tomorrow's world.

CHAPTER SIXTEEN

Journeys of Tomorrow

T HE YEARS HAD passed and my journey into the unfolding of stories and events "in search of truth" had been long and continuous. Sometimes arduous. The matching of the schematic of Von Braun's V2 rocket with the Tree of Life geometric drawing in Byrom's collection had been unexpected. It was as if I had stumbled upon it almost by accident, although I know I did not. I now found myself able to look forward to the future instead of continually looking at the past.

I often remind myself of what the Reverend Neville Barker Cryer said to me when my research led along an avenue that uncovered little-known 'nuggets' of fact: "Your duty is to record what you find. It is up to others to decide how they use it." I have asked myself many times on my journey as to whether I should follow that advice and confess that I have not always done so. More recent discoveries now tell me that there is another kind of knowledge which needs to be shared. It begins with the correspondence of another researcher, a story I now tell.

I had made arrangements for Dr. Barbera Romer, a German film producer based in New York and Miriam Dall' Igna, a systems analyst from the architects Foster and Partners, to spend the day with me here in Manchester on Tuesday, 13 September 2016. They were coming to study the Elizabethan theatre models that I had fostered creating, models made from the drawings (plans) in the Byrom Collection, including the Globe, The Rose, The Hope, The Swan, etc., assembled into wooden 3D models. Sir Mark Rylance of Globe theatre fame had suggested that Dr. Romer should contact me to take a look.

A few days before, on Friday 9 September at 5:00PM, I received a telephone call from a member of the staff at my literary agent's office in London. The purpose of the call was to inform me that they had received a parcel from America together with a letter addressed to me. I thought it might have something to do with my pending American visitors coming the following Tuesday. I asked the office person to open it because of the nearness of the anticipated visit. The contents were not anything to do with my New York visitors so I requested that they forward the parcel and letter.

The next few days were busy preparing for my expected visitors and exhibition. I did not have time to think of the mysterious parcel on its way from another American, via London to Manchester. Eventually it arrived. The parcel contained a folder of papers, professionally typed, artistically presented, and the letter. The letter-writer wished to be known simply as a 'carpenter,' a 'mason' in the trade[1]. He was addressing me, a successful author, with whom he would like to participate in a joint venture. The prospect seemed rather eccentric at the time. In the accompanying letter he said that I was "someone he had known in another life and never let go." There was a confident assertion as to my capabilities which I thought was rather presumptuous, and while I was polite and acknowledged the interest, I detached myself from further correspondence. Despite that decision, from time to time I felt slightly uncomfortable about a missed opportunity, without giving the letter-writer a chance for clarification. I was extremely busy at the time.

To my surprise, four months later, on 19 January 2017, a phone call from my agent informed me that a 'gift' had been delivered to them by hand, again addressed to me. It was too large to be forwarded without careful 'packaging,' and since the office would be closed for a time, beginning the next day, the letter accompanying the gift would be forwarded and the package kept there until the office re-opened the following month.

The gift was from this same 'carpenter' who had written earlier. He had visited London that December, hoping to deliver the package to me in person, but Andrew Mann, my agent, had moved its offices and he was unable to locate the new address in time for his planned return to America. Still, a representative of his did deliver the package to the new office at the beginning of the New Year.

The contents of this, his second letter, again, I found disquieting. The letter-writer was stating that through my published work, by presenting this rare collection of geometric pieces once owned by John Byrom, he himself could identify an interpretation relevant to himself and others. Something important.

It was 33 days before I could collect this 'gift.'

On Tuesday 21 February 2017, I made a day trip to London by train to collect what had been intended for me earlier. I made the journey as quickly as I could and was back in Manchester by 6:30PM, relieved that the task had been accomplished. It had been a tiring day.

The gift was unwrapped the next day, studied and hung in the hallway of my home along with other artefacts which I keep there. It's a wooden frame of nine windows, (3x3) each containing a geometric representation of what he calls a cosmic 'mandala,' with different features highlighted within the geometry of each of the nine windows. The property-maintenance 'joiner' at my home positioned the frame on the wall. He agreed that it was very well crafted.

The carpenter stated in the accompanying letter that he wanted this 'frame' to remain at a 'Byrom Resource Base' – me. It had been hanging in the hallway of his home in America for years. My diary entry for 22 February 2017 indicates a little of the drama to come:[2]

> I showed David the nine windows. David said that the workmanship involved was impressive. I did send a 'thank-you' note to Tina (Director of 'Andrew Mann'). She replied mid-afternoon. She had obviously spent some time thinking through after I had left (yesterday) and saw potential!! I noticed how interested she was in the NASA book – I decided to order her a copy.

I sent a courtesy 'thank-you' letter to the carpenter and this time headed it with my Manchester address. After all, given that his present was hanging in my hall, it seemed proper to let him know where it was. He wrote back acknowledging the information. There was no follow-up. At the time, pressures of one sort or another were considerable. There were demanding priorities, both personal and professional. This was in early 2017.

* * *

Although the previously mentioned correspondence between the late Rev. Cryer in Britain and Brother Brent Morris in America had been completed by 2009, I did not start the work and make the connections with Von Braun and the schematic of the 'Rocket' until 2020. That only happened because I was double-checking facts to do with John Dee and Edward Kelly in the light of the new material which had emerged, the results of my investigation into the 'allegorical paintings' presented here. They tell a story that can be puzzled out like the lyrics of a song. I have found that knowing the historical background helps.

Perhaps at this point, it is opportune to compare my researches to a Bach Prelude and Fugue in format. (With my background in music, I often think in these terms.) The 'Prelude' can be likened to the principle 'Subject,' John Byrom. The Fugue then becomes the various voices which enter at different stages in the theme, all related, culminating in a triumphant climax. So it seems that various links, as I have discovered them and as they have played out through my enquiries, were now being joined together.

It was on Wednesday, 25 September 2019, that this 'voice' from the past reappeared. It had been two and a half years since I had hung the gift from America in my hallway and acknowledged the generous gesture.

There had been no further contact between carpenter and me since then.

Using the old cliché that 'truth is stranger than fiction,' it came about like this: I had been invited by Claire Nahmad, who had reviewed my book "The Hidden Chapter" for the Journal 'Caduceus,' to participate in a ceremonial at the historic site of Rosslyn Chapel in Scotland earlier in the year and for various reasons, despite a number of discussions, the planned event never took place. Connections were made between group members and eventually it seemed quite natural for the 'carpenter' to re-open communication again, he having been miraculously turned up through Claire's efforts. It was all in a spirit of goodwill, shared amongst a loosely associated group, which Claire seemed to be assembling or at least entertaining.

The carpenter and I began to have exchanges regarding the geometry of the 'Byrom Collection.' They carried on for some months. The carpenter made it clear that he had an 'empathic' recognition of the geometry there, and by December 2019 he requested an opportunity to come to Britain to study the geometry here with me. Although initial murmurings took place I was very much involved with my new manuscript and once 2020 became the established New Year it was not long before planning for such a visit was interrupted by the coronavirus – limited travel and 'lockdowns.'

The correspondence continued and by April of 2020 a mutual respect had developed regarding the mathematics and philosophy associated with the representations (and interpretations) which we discussed. Unfortunately, there was a serious lapse in communication during the latter part of April and the whole of May. Technical problems with computers, together with the limits universally put in place because of the pandemic, disrupted dialogue.

On 2 June I received a message from the carpenter saying that communications intended for me via email had been returned to the server. We had not at this stage communicated by telephone. Technical hitches cleared up, despite the limitations imposed by the lockdown, and we were able to continue a meaningful dialogue. It was around this time that my attention had been drawn to a spate of 'crop circles' that began to appear in Wiltshire and the South of England.

It is a phenomenon that I had never studied on a regular basis. Although I was aware of the mysterious accuracy of the geometric patterns and the similarity to some of the patterns in the Byrom collection, I was too busy with my other studies to be distracted by yet another mystery. At this time however, discussions with respected colleagues tempted me to make myself aware of the most recent crop circles as they began to appear.

It was with a mild degree of shock that a fresh 'circle' had appeared on 30 May, just four days earlier (as the carpenter was trying to contact me), the design of which was the yin yang philosophical symbol, an exact replica of one of the 'nine windows' of geometric patterns in the carpenter's gift to me. It had the same proportions, mathematically correct in its geometry and symmetry. I say correct, because the background geometry of the entire 'window,' also common to all the other 'windows,' dictate the exact proportions and symmetry. I would not choose to comment on it except that the crop circle appeared shortly before I received the message from carpenter saying he had been trying to contact me without success.

It was a coincidence, of course, which I mentioned to him when I was next in communication (by email). A little later, I was surprised to read that he was not sure if I understood or had received all that he had left with me in the gift of 'nine windows.' Would I take it down from its hanging position and examine the frame thoroughly? The frame had been hanging in my hall since February 2017. It was in July 2020 that I got this message, more than three years later.

I was perplexed. The next morning, before my expected work folk appeared, I gingerly took the frame down, took it apart as directed and found nothing. Slightly embarrassed, what was I to make of it? What should I have found?

At the end of the work shift of my two helpers, I asked them without explaining why, if they would take the frame down from the wall. I did not tell them why I was asking them to do so or that I had done exactly what I was asking them to do earlier the same morning. It was the workman, David, who had hung it up in the first place, who took it down. He took the frame apart as I had done and began to examine the lining card holding the drawings firmly in position. With a very sharp eye, he noticed that the card was in fact two pieces, extremely finely fastened together. Separating the two pieces at the top, he peeled away tiny strips of cellotape. Between the two cards, hidden from the naked eye, were eighteen pieces of A4 paper containing the explanatory notes of the nine geometric drawings.[3] We stood there silently for a moment. My work people left almost immediately, with a sympathetic eye in my direction. "You are going to have some explaining to do," said David.

The frame had been hanging there for three and a half years without me realising anything of the eighteen pages inside. The 'piece' was put in place at a time when pressures for me were such that the details contained within the frame had been overlooked. And the secret within the 'nine windows' had remained hidden.

It quickly became apparent that the newly discovered eighteen pages not only should be studied with the 'nine windows' frame of geometric illustrations but also alongside the correspondence received earlier in late 2016. So I found myself looking at the pages for the first time, almost

This
Crop Circle
appeared
POTTERS FIELD
4th August 2020
U.K.

Emails from "carpenter"
3rd August and 4th August 2020

"Have a good day Joy, as you
awake and find 'me', when
you're ready.

Patterns within. Proof …
you hit the nail on the head
… it hangs in your hallway."

Tree of Life crop circle Potter's Field 2020

four years later on Wednesday, 1 July 2020. I wrote to the carpenter immediately to tell him of the find. There must have been awareness in my responses over time which demonstrated to him that some of this information was being overlooked. It was. I had not seen it. From this newly acquired understanding, aspects of my own research project had to be looked at from a different perspective. The von Braun 'Rocket Schemata' reflected in the 'Tree of Life' geometric drawing had played a pivotal role in my thinking.

It was the title of the carpenter's first correspondence which became something of a preoccupation, having read the explanatory eighteen pages:

"Being One"
The Meaning of Enlightenment
by
One Who Rakes
The story of a tree

In the carpenter's correspondence he said that he had known me "in another life." He also said that he recognised me when being interviewed on a TV programme. But who was he? He had travelled from America with this 'gift' hoping to meet me. That did not happen.

* * *

Sometimes we would do well to stand still and take stock of the journeys we have made through life, with the realisation that all that has gone before is to help us approach tomorrow. I had reached such a moment. The Reverend Barker Cryer had certainly been a presence in my decision-making process for years, but there was now yet another crossroads to negotiate, in which 'his' presence was still a reality. Sadly, he is now dead, as is my husband Allan.

Around the time of the publication of my last book "The Hidden Chapter" in 2011, another book had taken its place in the public arena: "Inside the Priory of Sion" by Robert Howells. The general business surrounding my own publication ensured this the book was not brought to my attention or my table for some time, my attention being elsewhere. My cocoon-like existence, deliberate or otherwise did not know of it at the time. It was only sometime later when a correspondent wrote to me that I had received an honourable mention in the book, that I obtained a copy. The reference was in connection with the 'material' I expose, matters referenced in my earlier publications, already in the public domain. Because I consider it to be relevant at this stage, I include the quote in its entirety. It is in connection with the list of Grand Masters of the 'Priory of Sion': [4]

Some of those on the original published list, including Robert Fludd, Johann Valentin Andrea, Robert Boyle, Isaac Newton and Charles Radclyffe, had all either influenced or directly contributed to what became the Byrom Collection papers. These later appeared in *The Queen's Chameleon,* a biography of poet John Byrom (1692-1763) by Joy Hancox. Hancox had discovered an entire archive of designs and writings from this period, as well as a list of their names. As this list was not discovered until the 1980's, the compiler of the *Dossiers Secrets,* which was published in the 1950's could not have known of its existence. Hancox' discoveries showed that this group were in contact with one another. Of particular interest is the fact that two diagrams from two different personal collections of these men have identical mistakes. This is evidence of important symbols being copied and circulated amongst an intellectual elite with esoteric knowledge and interests. It is likely that they were Rosicrucians, and because Rosicrucianism is effectively a branch of the Priory of Sion (see preceding chapter) this may well explain the appearance of these men on the original Grand Master list published in the *Dossiers Secrets.*

The organisation referred to as the Priory of Sion has touched the imagination of serious scholars as well as scoffers and film makers of all persuasions for a number of years. There are certain names and initials of personalities related to it inscribed on some of the drawings in the Byrom collection. That they were there before the year 1732 is a fact. And that they are directly related to the 'brotherhood' or 'Priory' is another fact. I am glad that Robert Howells chose to include that understanding in his own publication and I thank him here for doing so.

His book "Inside the Priory of Sion" brought me up starkly to another reality, one that I did not come to learn about until later. After the Reverend Barker Cryer's death, I received a list from his widow of the positions he held and organisations he belonged to during his lifetime – very distinguished. I had requested this list after his death in May 2013 so I could properly acknowledge his contribution to my work in any future publication of mine. Mrs. Marjorie Cryer kindly provided me with a formally prepared list of his prestigious achievements, many of which I already knew about. But on my first perusal, there was one that stood out, one that the Reverend had never mentioned in all the twenty-three years of our meetings together. On the list is included that he was a member of:

The Order of the Secret Monitor

According to Robert Howells in his "Inside the Priory," on the frontispiece of the Priory of Sion founding documents is a logo, a link between that organisation and the Secret Monitor. I quote:

Sion Frontispiece. The design is linked to the badge of Masonic Order of the Secret Monitor.

So, it seems that the Rev. Cryer had direct links with the Priory of Sion. For whatever reason, he could not, or did not tell me. In reappraising the Rev. Cryer's contribution over the years to my own work, I recognised that he had drawn my attention to certain facts that I doubt I would have come across if he had not told me. An example is the painting of the Tintern Abbey allegory 'Stem of Jesse.' If, in turn, I had contributed to the Secret Monitor I would not know. The legacy the Rev. Cryer gave me before and after his death, as a result of meetings, papers, letters and books far outweigh any doubts that might have lingered. It was in reflecting on this final cameo in connection with Reverend Cryer that I decided to look again at the carpenter's gift of Nine Windows, together with the papers he had sent and our ongoing correspondence.

* * *

Carpenter's correspondence was of a different order to any I had received along the way. Insistent, and in some ways 'coded,' it reminded me of the correspondence I had received many years earlier from Lord Birkett, although Lord Birkett's was not directly concerned with 'sacred geometry,' as this was. In the carpenter's case we were carrying out a written dialogue by email in the hope that our exchanges would develop into one of mutual benefit. Von Braun's Schemata and its corresponding companion piece The Tree of Life convinced me that carpenter's interest was more than of a general kind, and serious. Already there had been the coincidental link between the Wiltshire crop circle of 30 May and one of the 'nine windows.' From that day, ever cautious in my deliberations, I began to monitor the appearance of any crop circle that appeared in the U.K. At the same time, I registered the content of the carpenter's emails.

As the 'crop circle season' progressed, my recording of them became almost routine. At the same time, I was gathering what I considered to be responsible explanations for the extraordinary detail and precision in the patterns imposing themselves quite suddenly, usually overnight, in the fields of southern England. The explanations for these appearances are many and varied, but there are three main answers. The first is that they are 'man-made,' the second, they are made by 'extra-terrestrials' from another planet, third explanation: 'do not know but they are not made by humans.' I came to realise that thousands of interested observers are following this same phenomenon. There seem to be no mistakes in their creation, meaning that once the stalks of corn are bent they can't be made to stand upright. Is this the work of earthly humans, prone to making mistakes?

CROP CIRCLES U.K. 2020 Season

First Appearance

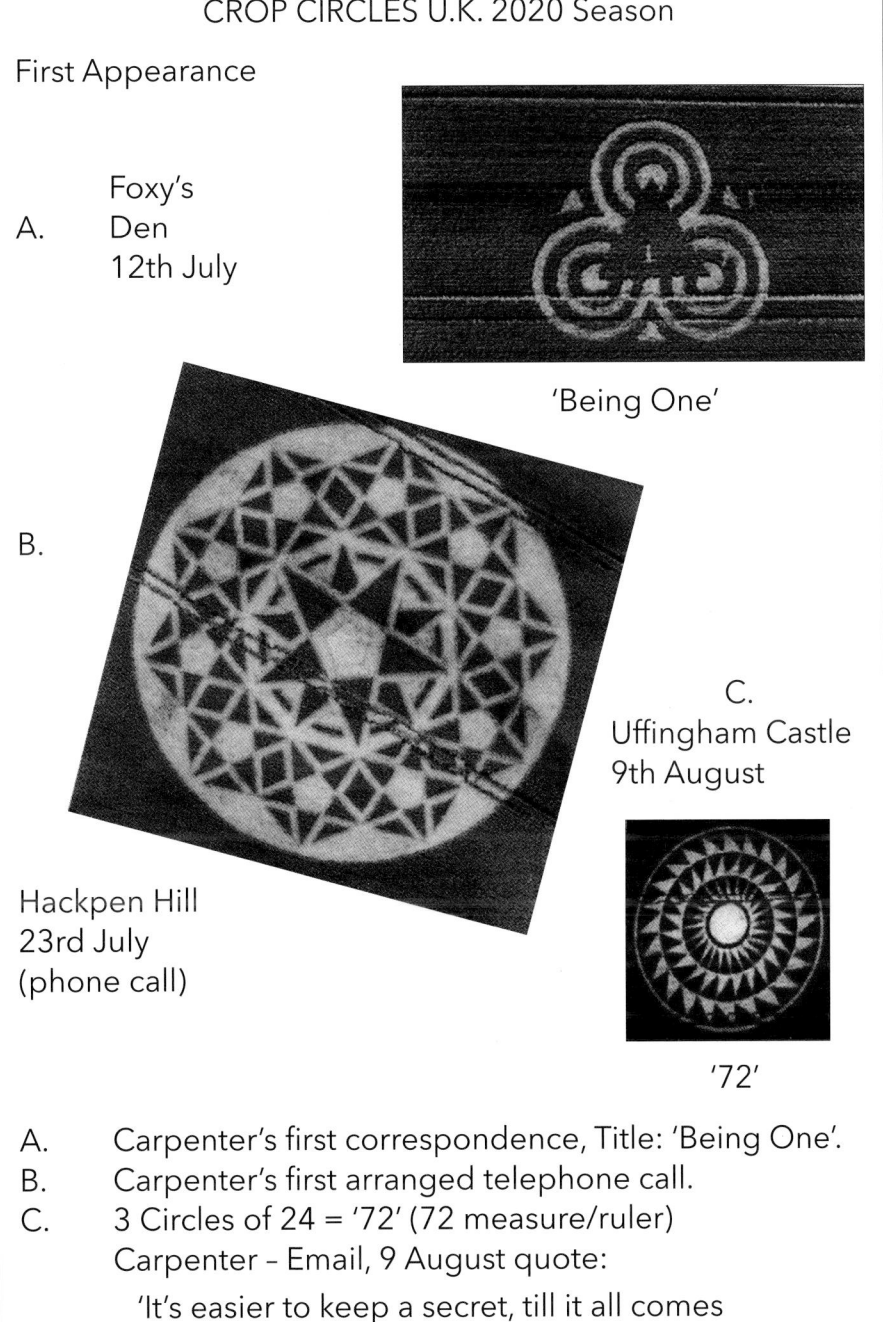

A. Foxy's Den 12th July

'Being One'

B.

C.
Uffingham Castle
9th August

Hackpen Hill
23rd July
(phone call)

'72'

A. Carpenter's first correspondence, Title: 'Being One'.
B. Carpenter's first arranged telephone call.
C. 3 Circles of 24 = '72' (72 measure/ruler)
   Carpenter – Email, 9 August quote:

   'It's easier to keep a secret, till it all comes
   out at once.'

Crop circles – Collage – 2021. Meaningful.

My curiosity became rather more serious. I came to notice a familiar detail here and there. I had by then reached a stage in my record-keeping where I told the carpenter I was keeping a dated record of the mysterious geometric patterns. We had by this time become more friendly (!) in our exchanges. We still had not had a meeting, but had agreed to discuss the possibility, amongst other topics on the telephone at 12:00 noon, Friday, 9 October. This was the second time we had a phone conversation. After our initial cordial greetings this time, we each embarked on our agenda, one of my questions was to ask 'carpenter' whether he was in any way connected with the appearance of the seasonal 'crop circles' as they became evident in the U.K. His answer was such that I could have asked more, but for prudential reasons decided not to. By the end of the long call it became evident that carpenter intended to visit Britain as soon as the conditions surrounding the coronavirus became more practical.

Later that same day, in keeping with good office practice, carpenter sent an email with a résumé of our telephone call. In this instance he chose to answer in more detail:[5]

> About the crop circles, I mean it when I say that they are speaking directly to you Joy. The point of the circles is to communicate a message to all who understand and you understand, you may be the only one who truly understands, which is why they are speaking directly to you. I mean specifically to you, as in "aliens to Joy, come in…over." They know what you and I are up to here; they planned it this way. They have spoken to me and now they speak to you, as you know. I am an emissary, as I have said, a fledgling blue in the world of aliens…I had to learn Galactic and to you they speak in symbolic logic, right down your geometrical alley. To top it all off, they know exactly what to say at exactly the right time, in exactly the right place for you to take notice. You've noticed that yourself, which is why it is so breath-taking, to know that you personally are being directly addressed by a real-life alien race. Me, I've been holding my breath for almost 50 years…

Later that day, I turned on my computer to look at the crop circle web site 'Stonehenge Dronescapes.' Although it had been reported that the 'crop circle season' was over, a repeat of an earlier crop circle appeared that day. It went under the title 'Hackpen Hill Crop Circle, Ghost Goes Yellow.' It had first appeared on 23 July, under the title 'Stunning Crop Circle.' This was the same day I had received my first telephone call from the carpenter. The same crop circle had therefore appeared on the same two dates as I had been engaged in telephone conversations with him. What's more: the principle of the 'design' of this now repeated crop circle was the same as that on the book jacket of "The Byrom Collection" hardback edition, in which the Tree of Life drawing for the schematic of the Rocket had appeared – see p. 219 of "The Byrom Collection."

"There are many seven-pointed stars in crop circles," carpenter noted. "Joy's star, Byrom's star, has eight, nigh-un to ten." (The carpenter says things like that, somewhat mysteriously.) By now, it seems to me that communications between this carpenter and me are known about somehow, in a mysterious but identifiable way, which can be recognised in this eerie crop circle. Someone who describes himself as an 'emissary' is by definition:

> A person sent as a diplomatic representative on a special mission.

He was then a very special messenger. But who is, or who are the Ring Masters of such proceedings? Certainly, I was being given summative information on this memorable day. Was I suffering from delusions? I don't think so. The evidence is in the public domain.

The official launch of my book "The Byrom Collection" was in May 1992. I have already included here the comment I made to those present at the time which I repeat: "Whilst there is one copy still in the bookshops, the knowledge within is available." It seems that a different 'phenomenon,' call it what you will, is ensuring that this knowledge is being confirmed in an unexpected format with a different delivery 'here in the flesh,' so to speak.

The Commissioning Editor Tom Maschler of Jonathan Cape Publishers, on the occasion of our agreement to publish my book "The Byrom Collection," suggested that I start a personal diary immediately. As is known to the reader, I have quoted several entries from those records in this narrative. Sadly, it must be confirmed that Tom Maschler died on 16 October, 2020, seven days after the crop circle with the uncanny central design appeared, the design identical to my book jacket, the same design that appeared in a crop circle in England on that 9 October, 2020, the day of the second telephone call, the same design that had appeared on 23 July, the day of the first phone call. Hmm. Call me a skeptic. It's been a journey of thirty years overall.

The carpenter intends to tell me, in person, some further details of his 'mission,' as he calls it. That meeting is to be as soon as he can make his promised visit from America to Manchester. In the meantime, my contemplative work continues in my self-imposed sanctuary, which itself has its own historical links with John Dee's habitat nearby, some likely here at my home. He lived for nine years (1595-1604) with his family at what is now Chetham's Library and Music School of Excellence here in

Manchester. The Chetham's Library is the oldest public reference library in the English speaking world and houses the collection of John Byrom's personal books and manuscripts as instructed by his great-grand daughter Eleanora Atherton in 1870. For me personally, my past experience teaching music and drama at that same school provides a poignant reminder of this historic building and what it has represented through the centuries since it was first built in 1412 to accommodate the priests of the Collegiate Church, now Manchester Cathedral. It seems that a philosophical 'circle' has been completed. One door has closed and another one is about to open.

I feel privileged to have had the opportunity to undertake such a journey. There have been times when I thought I had come to the end of this 'unfolding.' But then something might occur or be drawn to my attention and even at times a gift would be made, randomly it seemed, which would open yet another 'door' and I would step inside and find myself in strange surroundings which needed to be explored in the form of enquiry and learning, and understanding most importantly of all. I am about to begin the next stage of a journey I could never have imagined thirty years ago.

Prelude

INTERPRETING A SHEET of musical notation as a melody for an instrument or work for an orchestra has always held a fascination for me since early childhood. A script of a play, a score of an Opera or Musical as well as works of Art, as set pieces for the imagination to think about also have an immediate appeal. But 'codes' as used throughout the centuries by different cultures are a different order – testing the intelligence and the common sense of the human condition. The solving of the symbolism within particular codes I have found stimulating and gratifying.

Inquisitive, curious and 'enquiring' by nature, throughout my life's journey these qualities have provided me with wonders and delights. There have been sadnesses too, disappointments and on occasion frustrations and anger – all adding to life's rich pattern. It is called 'experience'!

As often happen, answers do not come easily and then a quest begins, at times challenging judgements. Since becoming the licensee of John Byrom's Collection of five hundred and sixteen geometric drawings many years ago, that has been the situation for me. The mystery of their purpose created with incredible precision and detail has been, and still is a preoccupation which has dictated the paths I have taken on my life's journey since 1984, when I first received them in two brown paper bags.

At times recognising my limitations, I have sought out those whose knowledge and skills were able to tell me what I needed to learn, thus taking me further on my journey. I have met, over the years some very wise individuals, anxious to share knowledge freely. I have also learned to respect the opinions of others. But the understanding of numerology as applied – within the geometry – requires a different recognition. It is orderly, compelling and mysterious. I was told fairly early in my study of the geometry, and before my first book "The Byrom Collection" was published in 1992 that the knowledge contained within particular examples could 'stop the holocaust'. At the time it was a premonition which I found bewildering and beyond my understanding. The remark became fixed in my psyche. Though bewildered – how should I proceed? My ambition became to seek out those who may have the knowledge to further my understanding and the project in positive ways.

My published books demonstrate the efforts made to use the information as and when discovered to that end. It was not always with the success that I hoped for, as will be understood. But the chosen paths have enriched my life beyond all expectations.

Always, during my quest over the years I have felt the support of a 'hidden hand' encouraging and leading me 'onwards'.

"The Messenger" is an account of filling in the gaps in the story of my journey over the years. Make of it what you will. Convinced by the existence of the 'hidden hand' throughout, I introduce whom I believe to be the real 'Messenger' in my journey thus far.

A Cadenza

A quote from Francis Bacon's Bi-Literal Cypher.

" I thinke some ray, that farre offe golden morning, will glimmer even into th' tombe where I shall lie, and I shall know that wisdome led me thus to wait unhonour'd, as is meete – rough hewe them how we will - doth even now knowe,- my justification bee complete.

"Farre off the day may be, yet in time here or hereafter it shall be understood. Though sorrowe is my constant companion now joy shall come on that morning.

"Though it shall not happen in mine owne day, this assurance that it cannot fail to come forth in due time, maketh weary labour less tiresome. It is noe doubt long to wait, but whatever should have been ordained by that Supreme Govenor of our lives doth give such satisfaction, it doth fully sustaine and succour th' heart, so that it surmounteth all fears."

The hidden hand

Coda

Until Francis Bacon/Sakespeare/Christ returns
The globe will remain as it is. I am hopeful. It
should be rebuilt.To that purpose I am dedicated.

Postscript
Yes, Joy.
I am asking you to say in your in your book I am here.
On Saturday, December 25,
2021 the Sun
will rise
in the
West in
Tintern.
It is
not by
my
Power
it is
done.

APPENDIX ONE

The Abbey Watergate at Tintern

Simpson   29th April 2021

UNTIL THE ARRIVAL of the Cistercian Monks following the Norman Invasion, the principal river crossing over the Wye in Tintern would have been the ford at Tintern Parva, mentioned in the Book of Llandaff, where Tewdric the Welsh king fought the Saxons losing his own life in that conflict. By the 12th Century the Monks were there with the Abbey and all the lands granted them by the Norman Conquerors. In order to access their land across the river they included the Watergate and slipway – the Abbey passage.

During the Monks stay in Tintern they ruled the roost with the blessing of the Norman tyrants, running the river as a fishery and obstructing the navigation on the lower tideway by constructing fish weirs which enabled them to net salmon on a regular basis as the fish migrated in runs up the river to spawn. The Wye being a salmon river was probably a serious consideration when deciding where to build the Abbey, I understand that the Cistercian Monks only ate fish and not meat. As a result of the weirs which obstructed the navigation they would have had to constantly maintain them because of the fast ebbs and flows of the river Wye and the Monks would have been on a confrontational footing with trading craft who would have been outside the Monks powerful jurisdiction with their right of navigation and needing to pass and repass with goods and commodities to the upper river ports of Brockweir, Monmouth, Ross and Hereford. All tidal rivers and creeks had a public right of navigation and were classed as arms of the sea from at least Saxon times.

I remember reading in a book by Keith Kissack that in 1301 Edward I tried to keep the river open and appointed a Royal Commission to inquire into the weirs between Hereford and Monmouth "as it appears that boats cannot pass as they were wont." Thirty years later the Lord of Monmouth was complaining that the Abbot of Tintern "had so heightened the weirs on the lower Wye that ships with wines and victuals cannot cross to the town of Monmouth and other places adjacent." The Tintern weirs were so important to the Monks because they needed to catch their fish.

It was always the lower Wye that caused the major disputes, since the meanders between Monmouth and Hereford have usually made it quicker to transport small goods overland. The Countess of Pembroke in 1296 had a barrel of venison delivered to her at Goodrich by cart after being brought from Bristol to Monmouth by boat.

In 1610 Rowland Vaughan of Bredwardine appealed to the Earl of Pembroke to demolish his weirs in the interest of the wine trade which was controlled from Chepstow Castle saying "good my lord down with the weirs let us have wine with our venison; the carriage of it from London by land makes a cup of claret look like a weak leane wench that hath the greene sickness and such as we have from Bristow is fitter to be drunke with a welch goate than an English buck."

There is an account of the Abbot at Tintern being in trouble with the Royal Commissioners' who had to keep coming to Tintern to throw down the weirs to allow boats, barges and lighters to pass as they are wont and no sooner was this done, they were built up again. Communications such as they were in those days would take time so this continued until Tudor times when things changed.

The effect of the weirs being built would have thrown the water back upstream and caused the ford to be submerged at low tide so the Abbot controlled the river crossing with a Watergate and ferryman's cottage in Tintern via the Abbey Passage ferry. The ferry continued throughout the ages belonging to the Dukes of Beaufort following the dissolution until the 10th Duke of Beaufort sold his interest in the ferry, with other property, in 1900. The old road from Chepstow via Porth Cassag, through the Abbey and the Watergate, over the river to Brockweir and then to Monmouth was replaced by the toll road in 1892. Picture records show that up until the time of the Turnpike Road, horses and carts were ferried across the river as well as foot passengers at the Abbey Passage. After the Turnpike Road was opened the large flat-bottomed double ended passing boat maintained by the Duke of Beaufort at his workshops at Pen-Y-Parc was replaced by a large rowing boat similar to a salmon stop net boat for carrying foot passengers.

A tour guide of 1910 shows the ferry still in use but there are no records to my knowledge of the ferry continuing after the First World War which was when the Railway Tramway Bridge in Tintern, owned by the Crown, could be used by the public. Ownership of the bridge passed in 1952 to the highway authorities jointly, Gloucestershire and Monmouthshire County Councils, to be kept open as a public carriage road.

Although the Abbey Passage Ferry was no longer in use the slipway continued to be used for commercial and pleasure purposes up until the 1960's.

During the Romantic period the Abbey Watergate slipway was a stopping off point for William Gilpin's "Wye Tour," William Wordsworth and his sister Dorothy crossed the river here and all the Wye tourists would have used the Abbey Passage slipways.

The Watergate and slipway are documented on the 1763 Badminton estate map held at the National Library of Wales in Aberystwyth and also in the sale catalogue of the Duke of Beaufort's estate of 1901. CADW still includes the Watergate, but not the slipway, in its area of protection today.

The proprietors of the Anchor and Beaufort Hotels filled in the slipway on the Monmouthshire bank in 1961 as the Beaufort Hotel was extending its car park and the slipway was a convenient place to tip the soil thus filling it in. They had hoped that this would stop the Anchor Inn from flooding but it didn't and we now realise that the part of the Anchor which flooded was in fact the boat house. During the 1960s the sewerage scheme added to the obstruction and since other services have compounded the situation. The slipway was still listed in 1972 as a launching site in the booklet "Getting Afloat" published by the Ship and Boat Builders National Federation.

A Public Inquiry took place in Tintern village Hall in January 2000 but although the Planning Inspectorate found that the slipway should be reopened nothing was done. Highway maps still show that the Ferry Road extends through the Watergate Archway down to the river.

James Tredinnick Simpson, 29th April 2021

Notes

Chapter One

1. Author. Daily Diary, 28 November, 1992.
2. Author. Daily Diary, 8 December, 1992.
3. Robert Fludd by Joselyn Godwin published 1979 by Thames and Hudson, London.
4. Theo Crosby. Original letter dated 8 January, 1993.

Chapter Two

1. Author. Daily Diary, 22 April, 1993.
2. Theo Crosby. Architectural Plans and correspondence dated June 1993 and June 1994, to Author with observations.
3. Theo Crosby. Original letter dated 10 May, 1994, (handwritten) to Author.
4. Author. Daily Diary, 15 April, 1993. A telephone call from H.M. I. Leon Crickmore – with information regarding Theo Crosby's death on 12 September.
5. Theo Crosby. Original letter dated 29 April, 1992 to Author.
6. Theo Crosby. Original letter dated 14 January, 1993 to Author.

Chapter Three

1. Author. Daily Diary, 23 January, 1995.
2. Lucy Beever. Original letter dated 25 April 1995. Public Relations Manager, Globe Theatre, London. To Author.
3. Author. Daily Diary, 5 January, 1996.
4. Professor Emeritus John Gleeson, San Francisco University. Original letter dated 9 May 1995, to Author.
5. Author. Daily Diary, 31 March, 1996.
6. Author. Daily Diary, 20 April, 1996.
7. Author. Daily Diary, 20 July, 1996.
8. Author. Daily Diary, 5 September, 1996.
9. Author. Daily Diary, 6 September, 1996.
10. Author. Daily Diary, 8 September, 1996.

11. Author. Daily Diary, 22 September, 1996.
12. Author. Daily Diary, 24 September, 1996.

Chapter Four

1. Francis Bacon - verse – the carpenter.
2. Author. Letter to Mark Rylance dated 19 October, 1996.
3. Mark Rylance. Letter dated 20 December, 1996, to Author.
4. Author. Daily Diary, 12 November, 1996.

Chapter Five

1. Author. Daily Diary, 28 April, 1997.
2. Author. Daily Diary, 29 April, 1997.
3. Author. Daily Diary, 11 June, 1997.
4. Author. Daily Diary. 12 June, 1997.
5. Author. Daily Diary, 14 June, 1997.
6. Author. Daily Diary, 15 June, 1997.
7. Author. Daily Diary, 18 June, 1997.
8. Author. Daily Diary, 26 June, 1997.
9. Author. Daily Diary, 7 August, 1997.
10. Author. Daily Diary, 21 August, 1997.
12. Author. Daily Diary, 25 July, 1997.
13. Author. Daily Diary, 30 August, 1997.

Chapter Six

1. Author. Daily Diary, 3 September, 1997.
2. Author. Daily Diary, 8 September, 1997.
3. Author. Daily Diary, 9 September, 1997.
4. Author. Daily Diary, 6 May, 1992.
5. Author. Daily Diary, 17 September, 1997.

Chapter Seven

1. Kingdom for a Stage published by Sutton Publishing Limited, Stroud, Gloucestershire 2001.
2. Lord Michael Birkett of Ulverston. Handwritten letter dated 28 February 2009.
3. The plaque remains in place as a recor of the industry operating in the area.

4. Recorded on Gloria Taylor's marriage papers.
5. Recorded on Glorias's marriage papers when she wed Michael Birkett.
6. Tom Maschler – Published Biography, Picador – 2005.

Chapter Eight

1. Author. Daily diary, 25 June, 2014.
2. Author. Daily Diary, 12 July, 2014.

Chapter Nine

1. 'Mr. Dyer' (Edward Dyer, 1543-1607) became godfather of John Dee's eldest son, Arthur two years later and was a regular visitor to Dee's household during his six years project in Europe.
2. Reference: A history of Monmouthshire, Sir Joseph Bradney (1913) Vol. 2. Part 2 P. 250. "Accurata gentle Herbertianae," which Sir Thomas had compiled. This is preserved in the Free Library at Cardiff and is a Folio MS book containing much valuable information, with beautifully executed water-colour drawings of tombs, and sketches of seals, etc. (The Author copied out the entry by hand of the 3rd Earl of Pembroke, Patron of Sir Thomas Herbert).
3. The Foreword was written by John Heron Lepper, P.G.D. Librarian and Curator to the United Grand Lodge of England. He describes the book as can 'justly be termed a handbook of masonic lore.'
4. Sir Thomas Herbert's 'Travels' book (1677). The Author acquired a copy published by 'EEBO Editions' which was a 'Reproduction of the original in the Henry E. Huntington Library and Art Gallery.'
5. Sir Henry Billingsley became Lord Mayor of London in 1596. He died in 1606. The genealogy of his family with close connections to the Tintern area can be seen in Sir Joseph Bradney's 'A History of Monmouthshire' Vol. 4. Part 2 Page 195.
6. 'The Nurtons.' Accumulative historic data, collated over the years suggests that this site was in use from the Roman period.
7. Reference: 'Servant of the Cecils, The Life of Sir Michael Hicks' by Alan G.R. Smith (1977) P. 148. Refers to the funeral of Robert Cecil, Lord High Treasurer who had died on 24 May 1612. Quote: 'The funeral, poorly attended, took place on June 9th. In the procession Hicks, carrying one of the banners traditional on such occasions, walked beside the dead man's cousin, Sir Francis Bacon – an honourable position. Just over two months later, on August 15th, he himself died.'

Chapter Ten

1. John Eglington Bailey, 1840-1888. His edition of John Dee's diary was published in 1880 (20 copies only). Edward Fenton. This edition was published in 1998 and is comprehensive of Dee's life – as a family man, an Elizabethan Magus and a man of letters. Bailey's edition is focused, as secretary of the Chetham's Society, and an antiquarian who was strict in his accuracy of original dates in his records but limited in his coverage.
2. There is finality about this diary entry which has to be in connection with Dee's visitors.
3. Tidder is another form of the word Tudor. Bacon regarded himself as a member of the royal house of Tudor. He wrote a history of Henry VII, the father of Henry VIII whom Bacon regarded as his great-grandfather.
4. Kelly was the 'scryer' (spiritual contact) for this 'exchange.' Dee and his son Arthur were present.
5. The Book. "The Burial of Francis Bacon and his mother in the Lichfield chapter House," by Walter Conrad Arensberg. Arensberg, 1878-1954 is responsible for the 'Francis Bacon Foundation Arensberg collection,' of 16,000 volumes housed in the Huntington Library in San Marino, America.
6. History. "City and Cathedral of Lichfield." Page 6 – reference to Robert Devereux (3rd Earl of Essex). Printed by Nichols and Son, Red Lion Passage, Fleet Street, London, 1805.
7. Isabel Holcroft. Ref. Text. "The secret history of the most renowned Queen Elizabeth and the Earl of Essex. By a person of quality." P. 24 – Quote: Well, Madam, since you will have it so, continues he, (Essex) I must acquaint you I am desperately in love with the Countess of Rutland; and that I cannot live if your Majesty consent not that she shall make me happy. (Isabel, 3rd countess had been widowed in 1587. Later in this book it alleges they were married secretly). Published by ECCO Literature and Language.

Chapter Eleven

1. The illustration itself had reminded the Reverend Cryer of another the Author had shown to him from the Schweighardt M.S. in the British Library. It was part of photographed papers that the Author had – which had not been included in earlier publications.
2. In the publication of the Collective Plays, the first Folio, in 1623, the play "The Tempest" is the first. Its introductory performance took

place in 1612 and was not the first play in the sequence as written. The Author considers its position as 'No. 1' to be significant.

3. Legends surrounding King Arthur, the Knights of the Round Table and Caerleon the Roman garrison which housed the second Augusta legion – give the whole area a mystical charm – which remains today. King Edward I – together with his immediate family members add to the mystique in the memory. His territorial dominance in medieval politics over much of Monmouthshire is on record.

4. By royal decree – 'The Edict of Expulsion,' King Edward I expelled 3,000 Jews from England in 1290. The edict was overturned by Oliver Cromwell in 1657.

Chapter Twelve

1. The Canals Studio had been a well-connected and established business, over a number of years, in Barcelona.

2. Manchester University had been given no indication of the Allegorical understanding by the Author, or of the illustration itself.

3. The Author thinks that Francis Bacon was a resident from 1626-1630, returning after the death of William Herbert, 3rd Earl of Pembroke.

4. "The Hidden Chapter" P.259 published by Byrom Projects 2011.

5. Alfred Dodd, op.cit, Vol 1, pp.80-1.

6. The Author was able to study the rare catalogue of the Nuremberg museum by Kurt Locher, Stuttgart 1997 S.306-309 at Manchester University Library.

7. The Nuremberg and Tintern Allegories were telling a story rooted in, and with the same tragic subject.

8. Ref. Priory of Sion Archives of Paul Smith. Vaincre No.3 September 1989, page 22 managing Editor: Thomas Plantard de Saint-Clair, 110, Rue Henri Dunant, 92700 Colombes.

Chapter Thirteen

1. The Author had no access to EBay sales and purchasing at the time.

2. "The Messianic Legacy" P.305

3. "The Messianic Legacy" P.305

4. Pierre de Cossé Brissac 1900-1993 (Duke of Brissac) wrote historical memoirs, four of which were about his family. He was the son of Francois de Cossé Brissac, the 11th Duke of Brissac from 1883-1944.

5. The Author Henry Lincoln discusses Jean Cocteau's altar fresco in the French Church, in his book "The Holy Place", published in 1991 by Jonathan Cape, London.

6. The Memoirs, privately published of the Sir Lambton Loraine family had been accessed by the Author at Manchester's Central Library.
7. John Loraine Baldwin was a personal friend of the 8th Duke of Beaufort.

Chapter Fourteen

1. Quote: "Views of Canterbury and Europe by L.L.Razé" (1805-1873) P.18, collated information by Brian Stewart. Published 1990 by Museums and Galleries, Canterbury.
2. The officially appointed cleric at St Michael's Church in 1850 was Rev. John Mais, but as Chaplain of Bristol Royal Infirmary, Rev. Thomas Tireman (The Nurtons) took most of the services – because of Mais's hospital commitments.
3. The links between the families of Audland, Tireman and Tintern were clearly of historic origins and Freemasonry.
4. Mr John Audland was officially appointed as medical practitioner in Tintern for over twenty years. His links with Freemasonry are clear.
5. A silver engraved communion chalice, engraved with King Charles 1st name, remains in use at St Michael's Church, Tintern.

Chapter Fifteen

1. Tony Colwell, the Editors remarks have been included as a matter of prudence.
2. "Transactions of Quatour Coronati Lodge No.2076". Volume 104 (1992) P.240 Review.
3. The Quatour Coronati Publication was printed in Great Britain by Butler & Tanner Ltd, Frome and London. This was the firm that 'lost' the negative of "The Byrom Collection". The firm closed in 2014.
4. Rev. Bill Stemper's enquiry reflects the interest from researchers with masonic backgrounds who recognised unusual aspects within the collection itself.
5. "The Hidden Chapter". Chapter - Some Reflections P.276.
6. Reverend Mr. Cryer – letter to Brother S. Brent Morris included a reference to the Author's enquiry quote:

'Steadily and consistently the intense study of these drawings has drawn her into matters, Masonic, both historical and philosophical, not excepting those dear to the Scottish rite'.

May I be allowed to suggest contact between you as I am sure there can be mutual benefit here. If you would like to make direct contact, her address is:_____.

The Author has had no direct contact with Brother, Brent Morris. He replied directly to Rev. Mr Cryer and addressed the queries that the Author had raised with him.

7. The Author does not question the validity of this entry.
8. This is a quote from a letter written by Bishop Godwin at a very sensitive time, within weeks and months of John Dee's death and the Bishop of Lichfield's demise 'Theodoricus' is/was a name with far reaching resonances – as already recorded.
9. The 'Bee' held highly symbolic meaning for the Merovingian dynasties – as well as for 300 bees covering the cloak Napoleon Bonaparte wore in 1804.
10. Rudolph Hess. Uncertain legacies – questions remain.

Chapter Sixteen

1. Mason. Not to be confused with Freemasonry membership.
2. The change of address for Andrew Mann Ltd. London, caused unexpected uncertainties for the American visitor at the time of his London stay.
3. It was so sudden. A new beginning happened that day.
4. "Inside the Priory of Sion" by Robert Howells, P.46, published by Watkins Publishing, London (2011).
5. The Author quotes this email in full, giving the statement the respect it deserves. After all, it was the protestant Bishop of Llandaff, Francis Godwin, (1562-1633) who wrote already cited "The Man in the Moone" first published in 1638. It is said by some critics to be one of the first works of science fiction and based on Francis Bacon's "Sylva Sylvarum" (1627). We are already facing reality. The journeys of tomorrow are being planned.

Acknowledgements

*T*HE *MESSENGER* has been in development for ten years. Duncan Beal the Managing Director of York Publishing Services Ltd. and his team of talented, skilled operators have been responsible for producing the finished book. His experience of the Publishing World at this particular time of change shone through the process of completion for which I am grateful. The technical competence and flair of the Design Manager Clare Brayshaw was in evidence from the beginning as was the sympathetic ear of Paula Charles, cradling, and in charge of publicity and promotion.

There are others that I must say 'thank you' to. My faithful four: David Preston, Dianne Johnson, Claire Lamb and Mary Routledge whose collective abilities saw to it that I had the necessary 'tools' alongside me at all times during periods of limited access due to the covid crisis. The same applies to my son Julian and his partner Stephanie who, despite their own busy professional commitments, found time to contribute their knowledge where I was lacking – in language translation and computer technology.

I must also recognise the continued support and interest of the Trustees of the Byrom Collection of geometric drawings as well as Sylvia and Aydin Ezen the owners of the Nuremberg Allegory and Madeline Gray whose surprise gift of the Razé Tintern View became so important to me.

There are those who are no longer with us whose contribution to the fabric of the story must not be forgotten. I salute them here. One of those was my husband Allan – always supportive it was his wish to be present at its completion. His contribution must be recorded along with the rest: Lord Michael Birkett, Brian Charles, Tony Colwell, Leon Crickmore, Theo Crosby, Rev. Neville Barker Cryer, John Davies, Prof. John Gleason, Tom Maschler, Dr. W. A. McCann, George Murcell, Claire Nahmad, Dr. Michael Powell, Bro. the Revd. William Stemper, Sir John Miles Huntington-Whiteley and Malcolm Young.

The contribution of members of the community of Tintern has been immeasurable. First of all the historical placement of 'The Nurtons' in the story is unique. The current owners Elsa and Adran Wood have been gracious in the sharing of this memorable site for the Launch of "The Hidden Chapter" and now "The Messenger". This note of appreciation includes Gemma Wood. Against the backdrop of Euroscan Ltd; Stratascan Ltd. specialising in Geophysics for Archaeology and Engineering and to

Cambrian Archaeological Projects Ltd, my thanks are due to 'The Wye Valley Hotel' who hosted innumerable events and meetings over the years on my behalf whilst carrying out my researches into St. Michael's church. The owners Sue and Barry Cook always give immensely gratifying service with aplomb and discretion. Members of St. Michael's Church administration have changed over recent years but I am indebted to the Rev. Nora Hill, Andrew Reid, Alan Carter and the current Warden Alan Hillard for their interest and support. A rich supply of anecdotal memorabilia has been freely given by Jim and Mary Simpson and Tony Hayward with serious local interest being shown by John Clarke and Jim Hewitt. I have also been fortunate in being able to avail myself of the wise counsel of Jo McCrum of the Author's Association on matters of publication.

I must also take the opportunity to thank the following whose variety of skills and expertise over the years have encouraged me to carry on when perhaps at times my energy levels were a little low. Suzie Hardie – over many years – a confidant, Michael and Margaret Darlington – legal advice and support, Elaine Ogden, Mark Ash, Angela Wood, Aline Watson – research assistants and supporters, Claire van Kampen, Peter Welsford – Bacon Society; Andrew Mann Ltd, Literary Agents; Ian Taylor T.F.E. Publishing, Toronto, Canada; Hiroshi Kato, LOE Associates, London; Laura Matthias and Lisa Wilson, 1604 Productions, USA; Dr. Barbera Romer, film producer, New York, USA; James Alan Egan, Author, Newport USA; The carpenter, USA – his gift of 'The Cosmic Mandala'.

My appreciation of Mark Rylance's initial interest at a critical time in my research programme – remains.

With regard to the illustrations used in the text, all have been acknowledged as to the Source. The Crop Circles appear by courtesy of "Stonehenge Dronescapes". With regard to the 'Tintern Allegory' I have used all reasonable endeavours to trace the copyright owner. Anyone claiming copyright should contact myself, the Author. This must also apply to any other legitimate claimant.

May I thank all named scholars and libraries whose work I have quoted from the sources listed. I learned so much along the route of my endeavours.

Finally, with complete admiration and trust I would also like to thank my editor, who believe you me, knows who he is.

Bibliography

Aldrin, Buzz, *Buzz Aldrin Magnificent Desolation* (Starr Buzz LLC-2009)

Arensberg, Walter Conrad, *The Burial of Francis Bacon and his mother in the Lichfield Chapter House)* (Kessinger Publishing Company, Montana, U.S.A.)

Arrowsmith, R.L., Hill, B.J.W., and Winlaw, A.S.R., *I Zingari* (JJG Publishing, 2006)

Attar, K.E., *Sir Edwin Durning-Lawrence: A Baconian & his books* (The Bibliographical Society 7th series, Vol.5, No.3, Sept.2004)

Aubrey, John, *Brief Lives*, ed. O.L. Dick (Penguin, 1987)

Audland, Sir Christopher, *Jenny* (Folio, Lancaster University, 2008)

Bacon, Francis, *The Advancement of Learning* (Dent, Everyman Library, 1954)

___ *Essays* (Oxford University Press, 1999)

Baigent, M., Leigh, R. and Lincoln, H., *The Holy Blood and The Holy Grail* (Corgi, 1983)

___*The Messianic Legacy (*Corgi, 1987)

Bailey, John Edlington, *Diary for the years 1595-1601* of Dr. John Dee *(Kessinger Publishing Co, U.S.A.)*

Beckh, Hermann, *Alchymy*, Temple Lodge, 2019

Benz, Portmann Izutsu et al, *Color Symbolism*, (Spring Publications, Inc. Dallas, Texas)

The Bible

Blatner, David, *The Joy of Pi* (Walker and Company, New York)

Bokenham, T.D., *Bacon, Shakespeare & the Rosicrucians* (privately printed, 1994)

Bosanquet Papers (Cwmbran Record Office)

Bowen, Catherine Drinker, *Francis Bacon, The Temper of the Man* (Hamish Hamilton, 1963)

Bradney, Sir Joseph, *A History of Monmouthshire*, 4 volumes (1904-33)

Brown, Allan and Michell John, *Crooked Soley Crop Circle* (The Squeeze Press, 2005)

Brown, Dan, *The Da Vinci Code* (Bantam Press, 2008)

Brown, Frederick, *An Impersonation of Angels* (Longmans, 1969)

Bunten, A. Chambers, *The Life of Alice Barnham* (Oliphant Ltd., 1928)

Byrom, John, *The Private Journal and Literacy Remains* (Chetham Society, Manchester, 1854)

Calendar of State Papers Domestic, 1611: 1626

Cats, Jacob, *Alle der Werken* (Amsterdam 1655 and 1658)

Censuses of England and Wales 1841-1991

Chaplin, Patrice, *Albany Park (Sceptre, 1987)*

_____ *Another City* (The Atlantic Monthly Press, New York, 1988)

_____*Happy Hour* (Pan Books, 1999)

_____*City of Secrets* (Robinson, 2007)

The Clergy Lists (published annually by Ecclesiastical Gazette Office, 19th century)

Clowes, W.B, *Family Business,*(William Clowes and Sons Ltd – 1969)

Clow, Barbara Hand and Gerry, *Alchemy of Nine Dimensions,* (Hampton Roads Publishing Company inc. 2004)

Coates, S.D., *The Water Powered Industries of the Lower Wye Valley* (Monmouth Borough Museums Service, 1992)

Cockburn, N.B., *The Bacon-Shakespeare Question* (privately printed, 1998)

Cocteau, Jean, *Past Tense, Diaries Volume One* (Hamish Hamilton, 1987)

_____*Past Tense, Diaries Volume Two* (Methuen, 1990)

Colvillle, Sir John, *Those Lambtons!* (Hodder & Stoughton, 1988)

Crockford's Clerical Directory (published annually)

Cryer, N.B., *The Arch and The Rainbow* (Lewis Masonic, 1996)

_____*Masonic Halls of England – The South* (Lewis Masonic, 1989)

_____*Masonic Halls of North Wales* (Lewis Masonic, 1990)

_____*Masonic Halls of South Wales* (Lewis Masonic, 1990)

Das Hausbuch der Mendelschen Zwolfbruderstiftung zu Nurnberg (Bruckmann, Munich, 1965)

Davies, Mrs. Andrew, *The History of the Parish of Carno* (Montgomery Collection Vol.33)

Davies, Robert, *Thomas Herbert* (The Yorkshire Archaeological & Topographical Journal, Vol. I, 1870)

Dawkins, Peter, *Arcadia* (The Francis Bacon Research Trust, 1988)

____*Dedication to the Light* (The Francis Bacon Research Trust, 1984)

____*The Shakespeare Enigma* (Polair Publishing, 2004)

Day, Barry, *This Wooden 'O'* (The Shakespeare Globe Trust, Oberon Book, 1996)

Dee, John, *Monas Hierglyphica*, (Guliel Silvius 1564)

Dictionary of National Biography (Oxford, 1917 and 2004)

Dictionary of Welsh Biography Down to 1940

Dixon, Piers, *Double Diploma* Hutchinson, 1968)

Dodd, Alfred, *Francis Bacon's Personal Life Story* (Vol.I, Kessinger Pub. Co., Montana, n.d and Vol.II, Century Hutchinson, 1986)

Donald, M.B., *Elizabethan Monopolies, The History of the Company of Mineral & Battery Works, 1568-1604* (Oliver & Boyd, 1961)

Durant, Horatia, *Raglan Castle* (The Starling Press, Newport, 1980)

____*The Somerset Sequence* (Newman Neame, 1951)

Encyclopedia Britannica (2002)

Fellows, Virginia M., *The Shakespeare Code* (Snow Mountain Press, USA, 2006)

Fenton, Edward (ed.), *The Diaries of John Dee* (Day Books, Charlebury, 1998)

Fratres Roseae Crucis, *Secret Shakespearean Seals* (Forgotten Books Classic Reprint Series)

French, Peter J., *John Dee* (Routledge & Kegan Paul – 1972)

Fuller, Jean Overton, *Sir Francis Bacon* (George Mann, 1994)

Gardner, Laurence, *Bloodline of the Holy Grail* (Element, 1996)

Geoffrey of Monmouth, *The History of the Kings of Britain* (trans. L. Thorpe) (Penguin Classics, 1986)

Gleason, John, Professor Emeritus, *Claes Visscher* (San Francisco University, unpublished refer. Joy Hancox)

Godwin, Francis (ed.), William Poole, *The Man in the Moone* (Broadview Editions, Ontario, 2009)

Gorst-William, Jessica, *Elizabeth, The Winter Queen* (Abelard, 1977)

Gray, Jonathan, *Dead Men's Secrets* (Teach Services, Inc. 2014)

Green, Miranda and Howell, Ray, *Celtic Wales,* (University of Wales Press, Cardiff 2000)

____*The Gwent County History* (University of Wales Press, Cardiff, 2004)

Guy, John R. And Smith, Ewart B., *Ancient Gwent Churches* (Starling Press, 1979)

Hall, Manly, P., *Orders of Universal Reformation* (Philosophical Research Society, Los Angeles, 1949)

Hammond, Fred, *The Truth About the Search at Chepstow* (Bacon Society, *Baconiana*, February issue, 1932)

Hancox, Joy, *The Byrom Collection* (Jonathan Cape, 1992)

____*The Queen's Chameleon* (Jonathan Cape, 1994)

____*Kingdom for a Stage* (Sutton Publishing, 2001)

Herbert, Thomas, *Some Years of Travels,* (EEBO Editions) (1677)

Herbert, Sir Thomas, *Memoirs of the Last Years of King Charles* (Neame, 1813)

____*History of the Race of Herbert* (MS *Herbertorium Prosapia c. 1651,* Cardiff Central

Library)

Hess, Wolf Rüdiger, *My Father Rudolph Hess* (Star Books 1987)

Higgins, Frank C., *Ancient Freemasonary* (Pyramid Books, New York, 1923)

Hoagland C. Richard – Bara, Mike, *Dark Mission,* (Feral House – 1009 Los Angeles, CA 90039)

Howells, Robert, *Inside Ike Priory of Sion,* (Watkins Publishing 2011)

Hyde, Montgomery, *Norman Birkett,* (Hamish Hamilton – 1965)

Irving, David, *Hess, The Missing Years 1941-1945,* Macmillan, London 1987)

Irving, Rob, Lundberg, John, *The Field Guide, The Art, History and Philosphy of Crop Circle making)* (Strange, Attractor Press, Bmsap, London, WC1N 3XX)

Jackson, John, *History of the City and Cathedral of Lichfield,* (Nichols and Son, 1805)

Jardine, Lisa and Stewart, Alan, *Hostage to Fortune: The Troubled Life of Francis Bacon (*Victor Gollancz, 1998)

Johnson, Edward, *Bacon-Shakespeare Coincidences* (The Bacon Society, 1950)

____*Francis Bacon's Maze* (Francis Bacon Research Society, 1961)

Jones, Bernard E., *Freemasons' Guide and Compendium* (Harrap, 1950)

Jones, Judith, *Monmouthshire Wills* (South Wales Record Society, Cardiff, 1997)

Jordan, W.K., *The Threshold of Power* (Allen & Unwin, 1990)

Kenyon, John R., *Raglan Castle* (Revised edition, Cadw, 2003)

King, Sir Edwin and Luke, Sir Harry, *The Knights of St. John in the British Realm* (St. John's Gate, 1967)

King, T.W. and Raines, F.R., *Lancashire Funeral Certificates* (Chetham Society, Manchester, 1869)

Kissack, K.E., *Medieval Monmouth* (Monmouth Historical & Educational Trust, 1974)

Kopec, John, *The Sabines at Riverbank* (Acoustical Society of America, New York, 1997)

Lamy, Lucie, *Egyptian Mysteries* (Thames & Hudson, 1989)

Lee, Sir Sidney (ed.), *The Life of Edward, Lord Herbert of Cherbury* (1886)

Like a Tree Planted: Brockweir Moravian Church 1833-1933 (privately printed)

Lilien Otto, M., J. *Christofle Le Blon* (Anton Hiersemann, Stuttgart, 1985)

Lincoln, Henry, *The Holy |Place,* (Jonathan Cape, 1991)

Loraine, Sir Lambton, *Pedigree and Memoirs of the Family of Loraine of Kirkharle* (J.B. Nichols and Sons, 1902)

Maier, Michael, ed. J. Godwin, *Atalanta Fugiens* (Phanes Press, MI USA, 1989)

Marrs, Jim, *Rule by Secrecy* (Perennial, 2001)

_____ *Alien Agenda* (Harper Collins, 1997)

Matthews Street Directory of Bristol

Maschler, Tom, *Tom Maschler, Publisher* (Picador, 2005)

Melchizedek, Drunvalo, *The Ancient Secret of the Flower of Life* (Light Technology Publishing 1990)

McCann, W.A., *Reports of Geophysical Surveys, Nov. 2000 & July 2001; Borehole Investigation July 2001 at St Michael and All Angels, Tintern* (unpublished)

Michell, John, *The Dimensions of Paradise* (Harper & Row, San Francisco, 1986)

___*The Temple of Jerusalem: A Revelation* (Gothic Image Publications, Glastonbury, 2000)

___*Who Wrote Shakespeare?* (Thames & Hudson, 1996)

Minutes of the Infirmary Committee, Bristol (Original MS, Bristol Record Office)

Morris, S. Brent Ph.d, *Why Thirty-Three,* (West Phalia Press 2019)

Morton, Alan Q. and Wess, Jane, *Public and Private Science, The King George III Collection* (Oxford University Press and The Science Museum, 1993)

Munro Smith, G., *A History of Bristol Royal Infirmary* (J.W. Arrowsmith, 1917)

Nixon, J.A., *Bristol Royal Infirmary. The Association of Bristol Royal Infirmary with British Masonry* (T.&W. Goulding Ltd., Bristol, n.d.)

Nooks & Crannies of Old Monmouthshire. A Catalogue of Paintings by Mary Ellen Bagnall-Oakley, 1833-1904 (Monmouth Museum)

Oakeshott, R. Ewart, *The Sword in the Age of Chivalry* (Arms and Armour Press, 1964)

Overton Fuller, Jean, *Sir Francis Bacon* (George, Mann, Maidstone, 1994)

Owen, Orville, W., *Sir Francis Bacon's Cipher Story* (Kessinger Pub. Co., Montana, n. d.)

Paar, H.W. and Tucker, D.G, *The Technology of Wire-Making at Tintern, Gwent, 1566-c1880* (Historical Metallurgy, 11:1, 1977)

Person of Quality, *The secret history of the most renowned Queen Elizabeth and the Earl of Essex.* (Ecco Print Editions)

Puttnam, W.G., *Excavations at Caer Neddfa Carno* (Montgomery Collection, Vol. 60)

Rees, William, *Industry Before the Industrial Revolution* (The University of Wales, Cardiff, 2 vols, 1968)

Rex, Richard, *The Tudors* (Tempus Publishing Ltd., Stroud, 2002)

Reynolds, John Lawrence, *Secret Societies,* Arcade Publishing, New York 2006

Rickards, Robert *Church and Priory of St. Mary, Usk* (Bemrose and Sons Ltd., 1904)

Robinson, David, M. (ed.), *The Cistercian Abbeys of Britain* (Batsford, 1998)

Rohl, John C.G., Warren, Martin and Hunt, David, *Purple Secret* (Bantam Press 1998)

Rosenthal, Daniel, *The National Theatre Story* (Oberon Books, 2013)

Russell, Bertrand, *Autobiography of Bertrand Russell* (George Allen & Unwin, 3 vols, 1967-9)

Russell, Judith (ed.), *Tintern's Story* (P.C.C., St Michael's Church, Tintern, 1990)

Schweighardt, Theophilus, *Speculum Sophicum Rhodo-Stauroticum* (1618)

Selenus Gustavus, *Cryptomenytices et Cryptographiae,* Libri IX (1624)

Shakespeare, Nicholas, *Bruce Chatwin,* (Vintage, in association with Harvard 1999 Press)

Sigma, Rho, *Ether – Technology* (Rho Sigma. Publisher)

Silvester, R.J., *The Llanwddyn Hospitium* (Montgomery Collection, Vol. 85, 1947)

Sinclair, Andrew, *The Sword and the Grail* (Century, 1993)

Sprigge, Elizabeth – Kihm, Jean-Jacques, *Jean Cocteau: The Man in the Mirror* (Victor Gollancz Ltd. London, 1968)

Stow, John, *Survey of London* (Dent, Everyman's Library, 1995)

Tintern Guide Book (Ministry of Buildings & Works, 1965)

Treue, Wilhelm *et al.* (eds.), *Das Hausbuch der Mendelschen Zwolfbruderstiftung zu Nurnberg* (Munich, Brukmann, 1965)

Vaughan, Rowland, ed. E.b.Wood, *Most Approved & Long Experienced Waterworkes* (1610)

Victoria History of the County of Hertfordshire (St Catherine's Press, 1912-23)

Von Braun, Wernher, *The Third Book of Words* (Simon Schuster, 1962)

____*The Mars Project* (University of Illinois Press 1962)

Wallace-Hadrill, J.M., *The Long-Haired Kings* (Toronto University Press, 1962)

Ward, E. and Richard Blake, *The Royal Lodge of Bristol and its R.A. and K.T. Appendages* (Transactions of the Quatuor Coronati Lodge, 1960)

Waterfield, Gordon, *Professional Diplomat: Sir Percy Loraine of Kirkharle, Bt.* (John Murray, 1973)

Watson, George, *Militia and Sappers* (The Castle Regimental Museum, Monmouth, 1996)

Watson, Peter, *Landscape of Lies* (Hutchinson 1989)

Weir, Alison, *Eleanor of Aquitaine* (Pimlico, 2000)

___ *Traitors of the Tower* (Vintage Books 2010)

Westminster Abbey Official Guide, 1977 edition

Wheatley, Dennis, *The Deception Planners* (Hutchinson, 1980)

Wimpenny, J. ed.), *Trellech 2000* (Biolime, Cardiff, 2000)

Wood, David, *Genesis* (The Baton Press, 1985)

Woolley, Benjamin, *The Queen's Conjuror* (Harper Collins, 2001)

Yates, Frances, *The Occult Philosophy* (Routledge & Kegan Paul, 1983)

___*The Rosicrucian Enlightenment* (Ark, 1986)

___*Theatre of the World* (Routledge & Kegan Paul, 1969)

Index